Springer Proceedings in Mathematics & Statistics

Volume 19

Springer Proceedings in Mathematics & Statistics

This book series features volumes composed of select contributions from workshops and conferences in all areas of current research in mathematics and statistics, including OR and optimization. In addition to an overall evaluation of the interest, scientific quality, and timeliness of each proposal at the hands of the publisher, individual contributions are all refereed to the high quality standards of leading journals in the field. Thus, this series provides the research community with well-edited, authoritative reports on developments in the most exciting areas of mathematical and statistical research today.

Mark Cummins • Finbarr Murphy • John J.H. Miller
Editors

Topics in Numerical Methods for Finance

 Springer

Editors
Mark Cummins
Dublin City University Business School
Dublin City University
Dublin 9, Ireland

Finbarr Murphy
Kemmy Business School
University of Limerick
Limerick, Ireland

John J.H. Miller
Institute for Numerical Computation
    and Analysis
7-9 Dame Court
Dublin 2, Ireland

ISSN 2194-1009   ISSN 2194-1017 (electronic)
ISBN 978-1-4899-7355-9   ISBN 978-1-4614-3433-7 (eBook)
DOI 10.1007/978-1-4614-3433-7
Springer New York Heidelberg Dordrecht London

Mathematics Subject Classification (2010): MSC 65

Printed on acid-free paper

Springer is part of Springer Science+Business Media (www.springer.com)

KEMMY
BUSINESS SCHOOL
University of Limerick

INSTITUTE FOR NUMERICAL
COMPUTATION AND ANALYSIS

UNIVERSITY *of* LIMERICK

OLLSCOIL LUIMNIGH

Fáilte Ireland

National Tourism Development Authority

meet in ireland.com

Bord Gáis Energy
think beyond

First Derivatives plc

Northern Trust

Preface

The papers in this volume were presented at the Numerical Methods for Finance Conference 2011, which was held at the University of Limerick, Ireland. All of the papers, with the exception of those of the keynote speakers, have been subjected to a rigorous refereeing procedure by the Editorial Committee. There is one additional paper, which was presented and accepted for publication in, but was accidentally omitted from, the published proceedings of the 2006 conference.

The aim of the conference series Numerical Methods for Finance is to attract leading international researchers from both academia and industry to discuss new research advances in, and applications of, numerical methods relevant to the solution of real problems in finance. This is a topic of practical importance because many of the mathematical models in quantitative finance cannot be treated analytically, and therefore must be solved numerically. Frequently this requires intensive computation on large grids of computers. In some respects, the development of numerical methods has kept pace with the development of computing hardware; however, many complex and high-dimensional problems are beyond the scope of even the most powerful contemporary computer clusters. Therefore, new numerical algorithms are required, which are fast, accurate and efficient for such problems. A wide range of topics and applications are presented in this volume. These offer both academic and practitioner appeal, reflecting the broad scope of the conference.

The 2011 conference was held under the joint auspices of the Institute for Numerical Computation and Analysis, Dublin, and the Kemmy Business School, University of Limerick. It is a pleasure to thank all members of the various committees who helped with the onerous burdens placed on them by the local organisers. The vital and generous support of the sponsors is also acknowledged with much gratitude. The dedicated work of all reviewers in the pre-conference review process and the post-conference proceedings review process is greatly appreciated. Finally, it was the participants who made this conference a lively, friendly and technically stimulating event. Particular thanks are extended to the keynote speakers who encouraged and facilitated the fascinating discussions and debates that emerged. It is to be hoped that participants will return to future conferences in the series.

Dublin, Ireland   Mark Cummins
Dublin, Ireland   John J. H. Miller
Limerick, Ireland   Finbarr Murphy

About the Editors

Mark Cummins is a Lecturer in Finance at the Dublin City University Business School. He holds a PhD in Quantitative Finance, with specialism in the application of integral transforms and the fast Fourier transform (FFT) for derivatives valuation and risk management. Mark has previous industry experience working as a Quantitative Analyst within the Global Risk function for BP Oil International Ltd., London. Mark has a keen interest in a broad range of energy modelling, derivatives, risk management and trading topics. He also has a growing interest in the area of sustainable energy finance, with particular focus on the carbon markets. Linked to Mark's industry experience, he holds a further interest in the area of model risk and model validation.

Finbarr Murphy is a Lecturer in Quantitative Finance at the University of Limerick, Ireland. Finbarr's key teaching and research interests lie in the field of credit risk and derivatives and more recently, in carbon finance. His research is focused on the application of generalised Lévy Processes and their application in the pricing and risk management of derivative products. Finbarr is also interested in the application of econometric techniques in finance. Prior to taking up his position in UL, Finbarr was a Vice President of Convertible Bond Trading with Merrill Lynch London.

John J.H. Miller is Director of INCA, the Institute for Numerical Computation and Analysis, in Dublin, Ireland. He is also a Fellow Emeritus of Trinity College, Dublin, where he was a member of the Mathematics Department. He received his Sc.D. from the University of Dublin and his Ph.D. in numerical analysis from the Massachusetts Institute of Technology. He completed his undergraduate degrees at Trinity College Dublin.

Contents

On Weak Predictor–Corrector Schemes for Jump-Diffusion Processes in Finance

Nicola Bruti-Liberati[†] and Eckhard Platen

Abstract Event-driven uncertainties such as corporate defaults, operational failures, or central bank announcements are important elements in the modeling of financial quantities. Therefore, stochastic differential equations (SDEs) of jump-diffusion type are often used in finance. We consider in this paper weak discrete time approximations of jump-diffusion SDEs which are appropriate for problems such as derivative pricing and the evaluation of risk measures. We present regular and jump-adapted predictor–corrector schemes with first and second order of weak convergence. The regular schemes are constructed on regular time discretizations that do not include jump times, while the jump-adapted schemes are based on time discretizations that include all jump times. A numerical analysis of the accuracy of these schemes when applied to the jump-diffusion Merton model is provided.

1 Introduction

Several empirical studies indicate that the dynamics of financial quantities exhibit jumps, see [2, 7, 16, 17]. Announcements by central banks, for instance, create jumps in the evolution of interest rates. Moreover, events such as corporate defaults and operational failures have a strong impact on financial quantities. These events cannot be properly modeled by pure diffusion processes. Therefore, several financial models are specified in terms of jump diffusions via their corresponding stochastic differential equations (SDEs), see [3, 8, 10, 20, 23].

[†]Died tragically in a traffic accident on his way to work.

N. Bruti-Liberati
School of Finance & Economics, University of Technology, Sydney, Australia

E. Platen (✉)
School of Finance & Economics and Department of Mathematical Sciences, University of Technology, Sydney, PO Box 123, Broadway, NSW 2007, Australia
e-mail: eckhard.platen@uts.edu.au

M. Cummins et al. (eds.), *Topics in Numerical Methods for Finance*, Springer Proceedings in Mathematics & Statistics 19, DOI 10.1007/978-1-4614-3433-7_1,
© Springer Science+Business Media New York 2012

1

The class of jump-diffusion SDEs that admits explicit solutions is rather limited. Therefore, it is important to develop discrete time approximations. An important application of these methods arises in the pricing and hedging of interest rate derivatives under the LIBOR market model. Since the arbitrage-free dynamics of the LIBOR rates are specified by nonlinear multidimensional SDEs, Monte Carlo simulation with discrete time approximations is the typical technique used for pricing and hedging. Recently, LIBOR market models with jumps have appeared in the literature, see [10, 28]. Here, efficient schemes for SDEs with jumps are needed.

Discrete time approximations of SDEs can be divided into the classes of strong and weak schemes. In the current paper, we study weak schemes which provide an approximation of the probability measure and are suitable for problems such as derivative pricing, the evaluation of moments, risk measures, and expected utilities. Strong schemes, instead, provide pathwise approximations which are appropriate for scenario simulation, filtering, and hedge simulation, see [19, 27].

A discrete time approximation Y^Δ converges weakly with order β to X at time T, if for each $g \in \mathscr{C}_P^{2(\beta+1)}(\mathbb{R}^d, \mathbb{R})$ there exist a positive constant C, independent of Δ, and a positive and finite number $\Delta_0 > 0$, such that

$$\varepsilon_w(\Delta) := |E(g(X_T)) - E(g(Y_T^\Delta))| \leq C\Delta^\beta, \tag{1}$$

for each $\Delta \in (0, \Delta_0)$. Here, we denote by $\mathscr{C}_P^{2(\beta+1)}(\mathbb{R}^d, \mathbb{R})$ the space of $2(\beta+1)$ continuously differentiable functions which, together with their partial derivatives of order up to $2(\beta+1)$, have polynomial growth. This means that for any $g \in \mathscr{C}_P^{2(\beta+1)}(\mathbb{R}^d, \mathbb{R})$ there exist constants $K > 0$ and $r \in \{1, 2, \ldots\}$, depending on g, such that

$$|\partial_y^j g(y)| \leq K(1 + |y|^{2r}),$$

for all $y \in \mathbb{R}^d$ and any partial derivative $\partial_y^j g(y)$ of order $j \leq 2(\beta+1)$.

In the case of pure diffusion SDEs, there is a substantial body of research on discrete time approximations, see [19]. The literature on weak approximations of jump-diffusion SDEs, instead, is rather limited, see [11–13, 21, 22, 24]. In this paper, we propose several new weak predictor–corrector schemes for jump-diffusion SDEs with first and second order of weak convergence.

For pure diffusion SDEs arising in applications to LIBOR market models, specific weak predictor–corrector schemes have been proposed and analyzed in [15, 18]. These authors show that for the numerical approximation of the nonlinear dynamics of discrete forward rates, predictor–corrector schemes outperform the simpler Euler scheme and allow the use of a single time step within reasonable accuracy. The weak predictor–corrector schemes proposed in the current paper can be applied to pricing and hedging of complex interest rate derivatives under LIBOR market models with jumps.

The paper is organized as follows. Sect. 2 introduces the class of jump-diffusion SDEs under consideration. In Sect. 3, we propose several weak predictor–corrector schemes for SDEs with jumps. These are divided into regular predictor–corrector schemes and jump-adapted predictor–corrector schemes. Finally, we present in Sect. 4 a numerical study of these schemes applied to the jump-diffusion Merton model.

2 Model Dynamics

The continuous uncertainty is modeled with an $\mathscr{A}$-adapted m-dimensional standard Wiener process denoted by $W = \{W_t = (W_t^1, \ldots, W_t^m)^\top, t \in [0,T]\}$, while the event-driven uncertainty is represented by an $\mathscr{A}$-adapted r-dimensional compound Poisson process denoted by $J = \{J_t = (J_t^1, \ldots, J_t^r)^\top, t \in [0,T]\}$. Each component J_t^k, for $k \in \{1,2,\ldots,r\}$, of the r-dimensional compound Poisson process $J = \{J_t = (J_t^1, \ldots, J_t^r)^\top, t \in [0,T]\}$ is defined by

$$J_t^k = \sum_{i=1}^{N_t^k} \xi_i^k,$$

where $N^1, \ldots, N^r$ are r independent standard Poisson processes with constant intensities $\lambda^1, \ldots, \lambda^r$, respectively. Let us note that each component of the compound Poisson process J^k generates a sequence of pairs $\{(\tau_i^k, \xi_i^k), i \in \{1,2,\ldots,N_T^k\}\}$ of jump times and marks. We will denote with $F^k(\cdot)$ the distribution function of the marks ξ_i^k, for $i \in \{1,2,\ldots,N_T^k\}$, generated by the kth Poisson process N^k.

We consider the dynamics of the underlying d-dimensional factors specified with the jump-diffusion SDE

$$dX_t = a(t,X_t)dt + b(t,X_t)dW_t + c(t,X_{t-})dJ_t, \tag{2}$$

for $t \in [0,T]$, with $X_0 \in \mathbb{R}^d$. Here $a(t,x)$ is a d-dimensional vector of real-valued functions on $[0,T] \times \mathbb{R}^d$, while $b(t,x)$ and $c(t,x)$ are a $d \times m$-matrix of real-valued functions on $[0,T] \times \mathbb{R}^d$ and a $d \times r$-matrix of real-valued functions on $[0,T] \times \mathbb{R}^d$, respectively. Moreover, we denote by $Z_{t-} = \lim_{s \uparrow t} Z_s$ the almost sure left-hand limit of $Z = \{Z_s, s \in [0,T]\}$ at time t. Let us note that in the following we adopt a superscript to denote vector components, which means, for instance, $a = (a^1, \ldots, a^d)^\top$. Moreover, we write b^i and c^i to denote the ith column of matrixes b and c, respectively.

We assume that the coefficient functions a, b, and c satisfy the usual linear growth and Lipschitz conditions sufficient for the existence and uniqueness of a strong solution of Eq. (2), see [25]. Moreover, when we will indicate the orders of weak convergence of the approximations to be presented in Sect. 3 we will assume that smoothness and integrability conditions similar to those required in [19] for pure diffusion SDEs are satisfied. The specific conditions along with a proof of the convergence theorem will be given in forthcoming work.

If we choose multiplicative coefficients in the one-dimensional case with one Wiener and one Poisson process, $d = m = r = 1$, then we obtain the SDE

$$dX_t = X_{t-} \left(\mu dt + \sigma dW_t + dJ_t \right), \tag{3}$$

which describes the *jump-diffusion Merton model*, see [23]. For this linear SDE, we have the explicit solution

$$X_t = X_0 e^{(\mu - \frac{1}{2}\sigma^2)t + \sigma W_t} \prod_{i=1}^{N_t} (1 + \xi_i), \tag{4}$$

which we will use in Sect. 4 for a numerical study. In [23] $(1 + \xi_i) = e^{\zeta_i}$ is the ith outcome of a log-normal random variable with $\zeta_i \sim \mathcal{N}(\rho, \varsigma)$. If instead $(1 + \xi_i)$ is drawn from a log-Laplace random variable we recover the *Kou model*, see [20]. Moreover, a simple degenerate case arises when $(1 + \xi_i)$ is a positive constant.

Other important examples of jump-diffusion dynamics of the form Eq. (2) arise in LIBOR market models. [28], for instance, consider a LIBOR market model with jumps for pricing short-term interest rate derivatives. Given a set of equidistant tenor dates $T_1, \ldots, T_{d+1}$, with $T_{i+1} - T_i = \delta$ for $i \in \{1, \ldots, d\}$, the components of the vector $X_t = (X_t^1, \ldots, X_t^d)^\top$ represent discrete forward rates at time t maturing at tenor dates $T_1, \ldots, T_d$, respectively. Moreover, they consider one driving Wiener process, $m = 1$, and two driving Poisson processes, $r = 2$. The diffusion coefficient is specified as $b(t,x) = \sigma x$, with σ a d-dimensional vector of positive numbers, and the jump coefficient $c(t,x) = \beta x$, where β is a $d \times 2$-matrix with $\beta^{i,1} > 0$ and $\beta^{i,2} < 0$, for $i \in \{1, \ldots, d\}$. In this way the first jump process generates upward jumps, while the second jump process creates downward jumps. Moreover, the marks are set to $\xi_i = 1$ so that the two driving jump processes are standard Poisson processes. A no-arbitrage restriction on the evolution of forward rates under the T_{d+1}-forward measure, see [3] and [10], imposes a particular form on the nonlinear drift coefficient $a(t,x)$ whose ith component is given by

$$a^i(t,x) = - \left\{ \sum_{j=i+1}^{d} \frac{\delta x^j}{1 + \delta x^j} \sigma^j + \lambda^1 \prod_{j=i+1}^{d} \left(1 + \beta^{j,1} \frac{\delta x^j}{1 + \delta x^j} \right) \right.$$
$$\left. + \lambda^2 \prod_{j=i+1}^{d} \left(1 + \beta^{j,2} \frac{\delta x^j}{1 + \delta x^j} \right) \right\}. \tag{5}$$

A complex nonlinear drift coefficient, as that in Eq. (5), is a typical feature of LIBOR market models. Therefore, it makes the application of numerical techniques essential in the pricing of complex interest rate derivatives.

To recover some empirical features observed in the market, it is sometimes important to consider a jump behavior more general than that driving the SDE, Eq. (2). By considering jump-diffusion SDEs driven by a Poisson random measure it is possible to introduce, for instance, state-dependent intensities. The numerical schemes to be presented can be naturally extended to the case with Poisson random measures.

3   Weak Predictor–Corrector Schemes

In this section, we present several discrete time weak approximations of the jump-diffusion SDE, Eq. (2). First, we consider regular schemes based on regular time discretizations which do not include jump times of the Poisson processes. Then we present jump-adapted schemes constructed on time discretizations which include all jump times.

3.1   Regular Weak Predictor–Corrector Schemes

We consider an equidistant time discretization $0 = t_0 < t_1 < \cdots < t_{\bar{n}} = T$, with $t_n = n\Delta$ and step size $\Delta = \frac{T}{\bar{n}}$, for $n \in \{0, 1, \ldots, \bar{n}\}$ and $\bar{n} \in \{1, 2, \ldots\}$. We denote a corresponding discrete time approximation of the solution X of the SDE, Eq. (2), by $Y^{\Delta} = \{Y_n^{\Delta}, \ n \in \{0, 1, \ldots, \bar{n}\}\}$.

Before introducing advanced predictor–corrector schemes, we present the *Euler scheme* which is given by

$$Y_{n+1} = Y_n + a\Delta + \sum_{j=1}^{m} b^j \Delta W_n^j + \sum_{k=1}^{r} c^k \hat{\xi}_n^k \Delta p_n^k, \tag{6}$$

for $n \in \{0, 1, \ldots, \bar{n}-1\}$, with initial value $Y_0 = X_0$. For ease of notation, we omit here and in the following the dependence on time and state variables in the coefficients of the scheme, this means we simply write a for $a(t_n, Y_n)$, etc.

In Eq. (6) we denote by $\Delta W_n^j = W_{t_{n+1}}^j - W_{t_n}^j \sim \mathcal{N}(0, \Delta)$ the nth increment of the jth Wiener process W^j and by $\Delta p_n^k = N_{t_{n+1}}^k - N_{t_n}^k \sim \text{Poiss}(\lambda^k \Delta)$ the nth increment of the kth Poisson process N^k with intensity λ^k. Moreover, $\hat{\xi}_n^k$ is the nth independent outcome of a random variable with given probability distribution function $F^k(\cdot)$. The Euler scheme achieves, in general, weak order of convergence $\beta = 1$.

It is possible to replace the Gaussian and Poisson random variables ΔW_n^j and Δp_n^k with simpler multipoint distributed random variables that satisfy certain moment-matching conditions. For instance, if we use in Eq. (6) the two-point distributed random variables $\Delta \widehat{W}_n^j$ and $\Delta \hat{p}_n^k$, where

$$P(\Delta \widehat{W}_n^j = \pm\sqrt{\Delta}) = \frac{1}{2}, \tag{7}$$

for $j \in \{1, \ldots, m\}$, and

$$P\left(\Delta \hat{p}_n^k = \frac{1}{2}(1 + 2\lambda^k \Delta \pm \sqrt{1 + 4\lambda^k \Delta})\right) = \frac{1 + 4\lambda^k \Delta \mp \sqrt{1 + 4\lambda^k \Delta}}{2(1 + 4\lambda^k \Delta)}, \tag{8}$$

for $k \in \{1, \ldots, r\}$, then we obtain the *simplified Euler scheme* which still achieves weak order of convergence $\beta = 1$. Let us note that this scheme can be implemented in a highly efficient manner by resorting to random bit generators and hardware accelerators, as shown for pure diffusion SDEs in [5, 6].

As indicated in [14] for pure diffusion SDEs and in [12, 13] for jump-diffusion SDEs, explicit schemes have narrower regions of numerical stability than corresponding implicit schemes. For this reason, implicit schemes for diffusion and jump-diffusion SDEs have been proposed. Despite their better numerical stability properties, implicit schemes carry, in general, an additional computational burden since they usually require the solution of an algebraic equation at each time step. Therefore, in choosing between an explicit and an implicit scheme one faces a trade-off between computational efficiency and numerical stability.

Predictor–corrector schemes are designed to retain the numerical stability properties of similar implicit schemes, while avoiding the additional computational effort required for solving an algebraic equation in each time step. This is achieved with the following procedure implemented at each time step: at first, an explicit scheme is generated, the so-called predictor, and afterward a de facto implicit scheme is used as corrector. The corrector is made explicit by using the predicted value $\bar{Y}_{n+1}$, instead of Y_{n+1}. The orders of weak convergence of the predictor–corrector schemes to be presented can be obtained by applying the Wagner–Platen expansion for jump-diffusion SDEs, see [26]. We refer to [4, 27] for the weak convergence of explicit and implicit approximations for SDEs with jumps.

The *weak order one predictor–corrector scheme* has corrector

$$Y_{n+1} = Y_n + \frac{1}{2} \{a(t_{n+1}, \bar{Y}_{n+1}) + a\} \Delta + \sum_{j=1}^{m} b^j \Delta W_n^j + \sum_{k=1}^{r} c^k \hat{\xi}_n^k \Delta p_n^k, \qquad (9)$$

and predictor

$$\bar{Y}_{n+1} = Y_n + a\Delta + \sum_{j=1}^{m} b^j \Delta W_n^j + \sum_{k=1}^{r} c^k, \quad \hat{\xi}_n^k \Delta p_n^k. \qquad (10)$$

The predictor–corrector scheme, Eqs. (9)–(10), achieves first order of weak convergence. Also in this case, we can use the two-point distributed random variables Eqs. (7) and (8) without affecting the order of weak convergence of the scheme. Let us note that the difference $Z_{n+1} := \bar{Y}_{n+1} - Y_{n+1}$ between the predicted and the corrected value provides an indication of the local error. This can be used to implement more advanced schemes with step-size control based on Z_{n+1}.

A more general *family of weak order one predictor–corrector schemes* is given by the corrector

$$Y_{n+1} = Y_n + \{\theta \, \bar{a}(t_{n+1}, \bar{Y}_{n+1}) + (1-\theta) \bar{a}\} \Delta$$
$$+ \sum_{j=1}^{m} \{\eta \, b^j(t_{n+1}, \bar{Y}_{n+1}) + (1-\eta) b^j\} \Delta W_n^j + \sum_{k=1}^{r} c^k \hat{\xi}_n^k \Delta p_n^k, \qquad (11)$$

for $\theta, \eta \in [0,1]$, where

$$\bar{a} = a - \eta \sum_{j=1}^{m} \sum_{i=1}^{d} b^{i,j} \frac{\partial b^j}{\partial x^i}, \tag{12}$$

and the predictor is as in Eq. (10). Here, one can tune the degree of implicitness in the drift coefficient and in the diffusion coefficient by changing the parameters $\theta, \eta \in [0,1]$, respectively. Note that when the degree of implicitness η is different from zero, it is important to use bounded random variables as $\Delta \widehat{W}_n^j$ and $\Delta \hat{p}_n^k$ in an implicit scheme. These prevent the effect of possible divisions by zero in the algorithm, see [19]. For a predictor–corrector method, this can be computationally advantageous, but it is no longer required. One can still use the Gaussian and Poisson random variables, ΔW_n^j and Δp_n^k, in the above scheme as in Eq. (11).

By using the Wagner–Platen expansion for jump-diffusion SDEs, it is possible to derive higher-order regular weak predictor–corrector schemes. However, these schemes are quite complex as they involve the generation of multiple stochastic integrals with respect to time, Wiener processes, and Poisson processes.

3.2 Jump-Adapted Weak Predictor–Corrector Schemes

As introduced in [26], let us consider a *jump-adapted time discretization* $0 = t_0 < t_1 < \cdots < t_M = T$ constructed as follows. First, as in Sect. 3.1, we choose an equidistant time discretization $0 = \bar{t}_0 < \bar{t}_1 < \cdots < \bar{t}_{\bar{n}} = T$, with $\bar{t}_n = n\Delta$, for $n \in \{1, \ldots, \bar{n}\}$, and step size $\Delta = \frac{T}{\bar{n}}$. Then we simulate all jump times τ_i^k, for $i \in \{1, 2, \ldots, N_T^k\}$ and $k \in \{1, \ldots, r\}$, generated by the r Poisson processes, and superimpose these on the equidistant time discretization. The resulting jump-adapted time discretization includes all jump times τ_i^k of the r Poisson processes and all equidistant time points $\bar{t}_1, \ldots, \bar{t}_n$. Its maximum step size is then guaranteed to be not greater than $\Delta = \frac{T}{\bar{n}}$. Note that the number $M + 1$ of points in the jump-adapted time discretization is random and, thus, changes in each simulation. It equals the total number of jumps τ_i^k of the r Poisson processes plus $\bar{n} + 1$. Therefore, the average number of grid points and, thus, of operations of jump-adapted schemes is for large intensity almost proportional to the total intensity $\bar{\lambda} = \sum_{k=1}^{r} \lambda^k$, which is defined as the sum of the intensities of the r Poisson processes.

From now on for convenience, we use the notation $Y_{t_n} = Y_n$ and denote by $Y_{t_{n+1}-} = \lim_{s \uparrow t_{n+1}} Y_s$ the almost sure left-hand limit of Y at time t_{n+1}.

Within a jump-adapted time discretization, by construction jumps arise only at discretization times and we can separate the diffusion part of the dynamics from the jump part. Therefore, the *jump-adapted Euler scheme* is given by

$$Y_{t_{n+1}-} = Y_{t_n} + a\Delta_{t_n} + \sum_{j=1}^{m} b^j \Delta W_{t_n}^i, \tag{13}$$

and

$$Y_{t_{n+1}} = Y_{t_{n+1}-} + \sum_{k=1}^{r} c^k(t_{n+1}-, Y_{t_{n+1}-}) \Delta J_{t_{n+1}}^k, \tag{14}$$

for $n \in \{0, \ldots, M-1\}$, where $\Delta_{t_n} = t_{n+1} - t_n$ and $\Delta W_{t_n}^j = W_{t_{n+1}}^j - W_{t_n}^j \sim \mathcal{N}(0, \Delta_{t_n})$. Here, $\Delta J_{t_{n+1}}^k$ equals $\xi_{N_{t_{n+1}}^k}^k$ if t_{n+1} is a jump time of the kth Poisson process or zero otherwise. The solution X follows a diffusion process between discretization points and is approximated by Eq. (13). If we encounter a jump time as discretization time, then the jump impact is simulated by Eq. (14). The jump-adapted Euler scheme has first order of weak convergence. By replacing the Gaussian random variable $\Delta W_{t_n}^j$ in Eq. (13) with the two-point random variable

$$P(\Delta \widehat{W}_{t_n}^j = \pm \sqrt{\Delta_{t_n}}) = \frac{1}{2}, \tag{15}$$

for $j \in \{1, \ldots, m\}$, we obtain the *jump-adapted simplified Euler scheme* which still achieves first order of weak convergence.

The *jump-adapted weak order one predictor–corrector scheme* is given by the corrector

$$Y_{t_{n+1}-} = Y_{t_n} + \frac{1}{2} \left\{ a(t_{n+1}-, \bar{Y}_{t_{n+1}-}) + a \right\} \Delta + \sum_{j=1}^{m} b^j \Delta W_{t_n}^j, \tag{16}$$

the predictor

$$\bar{Y}_{t_{n+1}-} = Y_{t_n} + a\Delta_{t_n} + \sum_{j=1}^{m} b^j \Delta W_{t_n}^j, \tag{17}$$

and Eq. (14). This scheme achieves the same first order of weak convergence of the jump-adapted Euler scheme. Thanks to the quasi-implicitness in the drift it has, in general, better numerical stability properties. Also in this case, it is possible to replace the Gaussian random variables in Eqs. (16) and (17) with the two-point random variables in Eq. (15).

A more general *family of jump-adapted weak order one predictor–corrector schemes* is given by the corrector

$$Y_{t_{n+1}-} = Y_{t_n} + \left\{ \theta \bar{a}(t_{n+1}-, \bar{Y}_{t_{n+1}-}) + (1-\theta)\bar{a} \right\} \Delta$$

$$+ \sum_{j=1}^{m} \left\{ \eta \, b^j(t_{n+1}-, \bar{Y}_{t_{n+1}-}) + (1-\eta) b^j \right\} \Delta W_n^j, \tag{18}$$

for $\theta, \eta \in [0, 1]$. Here $\bar{a}$ is defined as in Eq. (12) and the predictor as in Eq. (17) again together with a relation as in Eq. (14). This scheme achieves in general first order of weak convergence. Also in this case one can use the two-point random variables as in Eq. (15).

Within the class of jump-adapted schemes, we can derive higher-order weak predictor–corrector schemes which do not involve multiple stochastic integrals with respect to the Poisson processes. By using a second order weak implicit scheme as corrector and a second order weak explicit scheme as predictor, we obtain the *jump-adapted weak order two predictor–corrector scheme*. It is given by the corrector

$$Y_{t_{n+1}-} = Y_{t_n} + \frac{1}{2}\left\{a(t_{n+1}-, \bar{Y}_{t_{n+1}-}) + a\right\}\Delta_{t_n} + \Psi_{t_n}, \tag{19}$$

with

$$\Psi_{t_n} = \sum_{j=1}^{m}\left\{b^j + \frac{1}{2}L^0 b^j \Delta_{t_n}\right\}\Delta W_{t_n}^j + \frac{1}{2}\sum_{j_1,j_2=1}^{m} L^{j_1} b^{j_2}\left(\Delta W_{t_n}^{j_1}\Delta W_{t_n}^{j_2} + V_{t_n}^{j_1,j_2}\right), \tag{20}$$

the predictor

$$\bar{Y}_{t_{n+1}-} = Y_{t_n} + a\Delta_{t_n} + \Psi_{t_n} + \frac{1}{2}L^0 a(\Delta_{t_n})^2 + \frac{1}{2}\sum_{j=1}^{m} L^j a \Delta W_{t_n}^j \Delta_{t_n}, \tag{21}$$

and a relation as in Eq. (14). The differential operator L^0 is defined by

$$L^0 := \frac{\partial}{\partial t} + \sum_{i=1}^{d} a^i(t,x)\frac{\partial}{\partial x^i} + \frac{1}{2}\sum_{i,k=1}^{d}\sum_{j=1}^{m} b^{i,j}(t,x)b^{k,j}(t,x)\frac{\partial^2}{\partial x^i \partial x^j}, \tag{22}$$

and the operator L^j by

$$L^j := \sum_{i=1}^{d} b^{i,j}(t,x)\frac{\partial}{\partial x^i}, \tag{23}$$

for $j \in \{1,\ldots,m\}$. The random variables $V_{t_n}^{j_1,j_2}$ are two-point distributed with

$$P(V_{t_n}^{j_1,j_2} = \pm\sqrt{\Delta_{t_n}}) = \frac{1}{2}, \tag{24}$$

for $j_2 \in \{1,\ldots,j_1 - 1\}$, where

$$V_{t_n}^{j_1,j_1} = -\Delta_{t_n}, \tag{25}$$

and

$$V_{t_n}^{j_1,j_2} = -V_{t_n}^{j_2,j_1} \tag{26}$$

for $j_2 \in \{j_1 + 1,\ldots,m\}$ and $j_1 \in \{1,\ldots,m\}$. The Gaussian random variable $\Delta W_{t_n}^k$ can be replaced by the three-point random variable $\Delta\widetilde{W}_{t_n}^k$ defined by

$$P(\Delta\widetilde{W}_{t_n}^k = \pm\sqrt{3\Delta_{t_n}}) = \frac{1}{6}, \qquad P(\Delta\widetilde{W}_{t_n}^k = 0) = \frac{2}{3}, \tag{27}$$

for $k \in \{1,\ldots,m\}$.

Let us finally remark that, although jump-adapted schemes are easier to derive and implement than corresponding regular schemes, their computational complexity is for large intensities almost proportional to the sum of the intensities of the driving Poisson processes. Therefore, while jump-adapted schemes should be in general preferred, in the approximation of SDEs driven by high-intensity Poisson processes regular schemes are normally more efficient.

4   Numerical Results

For illustration in this section, we present some numerical results obtained by applying some of the schemes described in Sect. 3 to the evaluation of a payoff function g of the solution X of Eq. (2) at a terminal time T. We discretize in time the dynamics of the solution X of the linear SDE, Eq. (3), with one of the schemes in Sect. 3 and perform a Monte Carlo simulation to estimate $E(g(X_T))$. Note that this numerical approximation generates a systematic error, resulting from the time discretization of X, and also a statistical error caused by the finite sample size used in the Monte Carlo simulation. In this paper, we address the problem of reducing the systematic error. Therefore, in our experiments the number of simulation paths is chosen so large that the statistical error becomes negligible when compared to the systematic error. For variance-reduction techniques, which reduce the statistical error, we refer to [9, 19].

The theorems which establish the weak order of convergence of the discrete time approximations presented in Sect. 3 require a certain degree of smoothness for the payoff function g. Recall that we assumed in Sect. 1 that $g \in \mathscr{C}_P^{2(\beta+1)}(\mathbb{R}^d, \mathbb{R})$, see also [19]. The same conditions are usually required for weak convergence in the case of pure diffusion SDEs. For results with non-smooth payoffs limited to the Euler scheme for pure diffusion SDEs, we refer to [1].

As particular example we consider the evaluation of the expectation of the non-smooth payoff of a call option $g(x) = (x - K)^+$, where K is the strike price, that means we evaluate $E((X_T - K)^+)$. Since we have modeled the dynamics of the security with the jump-diffusion Merton model, by using Eq. (4) we obtain a closed form solution given by

$$E((X_T - K)^+) = e^{\left(\mu + \lambda(E(\xi)-1)\right)T} \sum_{n=0}^{\infty} \frac{e^{-\lambda'T}(\lambda'T)^n}{n!} f_n, \tag{28}$$

where $\lambda' = \lambda(E(\xi) - 1)$. Here,

$$f_n = X_0 \mathscr{N}(d_{1,n}) - e^{-\mu_n T} K \mathscr{N}(d_{2,n}), \tag{29}$$

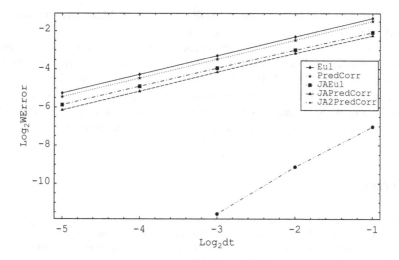

Fig. 1 Weak error for Euler, predictor–corrector, jump-adapted Euler, jump-adapted predictor–corrector, and jump-adapted weak order two predictor–corrector schemes

denotes the Black–Scholes price of a European call option with the parameters specified as

$$d_{1,n} = \frac{\ln(\frac{X_0}{K}) + (\mu_n + \frac{\sigma_n^2}{2})T}{\sigma_n \sqrt{T}}, \tag{30}$$

$d_{2,n} = d_{1,n} - \sigma_n \sqrt{T}$, $\mu_n = \mu - \lambda(E(\xi)) + \frac{n\ln\{E(\xi)-1\}}{T}$ and $\sigma_n^2 = \sigma^2 + \frac{n\varsigma}{T}$. We denote by $\mathcal{N}(\cdot)$ the probability distribution of a standard Gaussian random variable. Therefore, the weak error $\varepsilon_w(\Delta)$, defined in Eq. (1), can be explicitly computed.

In Fig. 1, we report a log–log plot with the logarithm $\log_2(\varepsilon_w(\Delta))$ of the weak error versus the logarithm $\log_2(\Delta)$ of the time step size. The parameters of the linear SDE, Eq. (3), are set as $\mu = 0.05$, $\sigma = 0.2$, and $\beta = 0.2$, with initial value $X_0 = 1$. The driving jump process is a standard Poisson process with intensity $\lambda = 0.2$. Therefore, the marks ξ are constant and equal one. Moreover, we chose the final time $T = 0.5$ and the strike price $K = 1.25$. We consider the Euler, the weak order one predictor–corrector, the jump-adapted Euler, the jump-adapted weak order one predictor–corrector and the jump-adapted weak order two predictor–corrector schemes. These are labeled "Eul," "PredCorr," "JAEuler," "JAPredCorr," and "JA2PredCorr," respectively, in Fig. 1. Note that in the log–log plot the achieved orders of convergence are given by the slopes of the observable lines. Figure 1 indicates that the Euler, the predictor–corrector, the jump-adapted Euler, and the jump-adapted predictor–corrector schemes achieve an order of weak convergence of about $\beta = 1$. By comparing the two Euler schemes to the corresponding predictor–corrector schemes, one notices that predictor–corrector schemes are more accurate. This effect is due to the implicitness in the drift of predictor–corrector schemes and is expected to be more pronounced for more complex nonlinear SDEs, which has been reported also in [15] for diffusion SDEs arising in BGM models. In particular, for stiff SDEs with widely varying time scales predictor–corrector schemes should be preferred to the traditional Euler scheme. When comparing the first-order regular schemes, "Eul" and "PredCorr" in the figure, to the first-order jump-adapted schemes, "JAEuler" and "JAPredCorr," we note that jump-adapted schemes are more accurate. This is due to the simulation of the jump impact at the correct jump times. Finally, the jump-adapted weak order two predictor–corrector scheme is the most accurate and seems to achieve second order of weak convergence.

References

1. Bally, V., Talay, D.: The law of the Euler scheme for stochastic differential equations I. Convergence rate of the distribution function. Probab. Theor. Relat. Field. **104**(1), 43–60 (1996)
2. Bates, D.: Jumps and stochastic volatility: exchange rate processes implicit in Deutschemark options. Rev. Financ. Stud. **9**(1), 96–107 (1996)
3. Björk, T., Kabannov, Y., Runggaldier, W.: Bond market structure in the presence of marked point processes. Math. Finance **7**, 211–239 (1997)
4. Bruti-Liberati, N.: Numerical solution of stochastic differential equations with jumps in finance. Ph.D. thesis, University of Technolology, Sydney, Australia (2007)
5. Bruti-Liberati, N., Platen, E.: On the efficiency of simplified weak Taylor schemes for Monte Carlo simulation in finance. In: Computational Science - ICCS 2004, Lecture Notes in Computer Science, vol. 3039, pp. 771–778. Springer, Berlin (2004)
6. Bruti-Liberati, N., Platen, E., Martini, F., Piccardi, M.: A multi-point distributed random variable accelerator for Monte Carlo simulation in finance. In: Proceedings of the Fifth International Conference on Intelligent Systems Design and Applications, pp. 532–537. IEEE Computer Society Press, Silver Spring, MD (2005)
7. Das, S.R.: The surprise element: jumps in interest rates. J. Econometrics **106**, 27–65 (2002)
8. Duffie, D., Pan, J., Singleton, K.: Transform analysis and option pricing for affine jump diffusions. Econometrica **68**, 1343–1376 (2000)
9. Glasserman, P.: Monte Carlo methods in financial engineering. Applications of Mathematics, vol. 53. Springer, Berlin (2004)
10. Glasserman, P., Kou, S.G.: The term structure of simple forward rates with jump risk. Math. Finance **13**(3), 383–410 (2003)
11. Glasserman, P., Merener, N.: Convergence of a discretization scheme for jump-diffusion processes with state-dependent intensities. Finance Stoch. **7**(1), 1–27 (2003)
12. Higham, D.J., Kloeden, P.E.: Numerical methods for nonlinear stochastic differential equations with jumps. Numer. Math. **110**(1), 101–119 (2005)
13. Higham, D.J., Kloeden, P.E.: Convergence and stability of implicit methods for jump-diffusion systems. Int. J. Numer. Anal. Model. **3**(2), 125–140 (2006)
14. Hofmann, N., Platen, E.: Stability of superimplicit numerical methods for stochastic differential equations. Fields Inst. Commun. **9**, 93–104 (1996)
15. Hunter, C.J., Jäckel, P., Joshi, M.S.: Getting the drift. Risk **14**(7), 81–84 (2001)
16. Johannes, M.: The statistical and economic role of jumps in continuous-time interest rate models. J. Finance **59**(1), 227–260 (2004)
17. Jorion, P.: On jump processes in the foreign exchange and stock markets. Rev. Financ. Stud. **1**, 427–445 (1988)
18. Joshi, M., Stacey, A.: New and robust drift approximations for the LIBOR market model. Quant. Finance **8**(4), 427–434 (2008)

19. Kloeden, P.E., Platen, E.: Numerical solution of stochastic differential equations. Applications of Mathematics, vol. 23. Springer, Berlin, Third Printing (1999)
20. Kou, S.G.: A jump diffusion model for option pricing. Manag. Sci. **48**, 1068–1101 (2002)
21. Kubilius, K., Platen, E.: Rate of weak convergence of the Euler approximation for diffusion processes with jumps. Monte Carlo Methods Appl. **8**(1), 83–96 (2002)
22. Liu, X.Q., Li, C.W.: Weak approximations and extrapolations of stochastic differential equations with jumps. SIAM J. Numer. Anal. **37**(6), 1747–1767 (2000)
23. Merton, R.C.: Option pricing when underlying stock returns are discontinuous. J. Finance Econ. **2**, 125–144 (1976)
24. Mikulevicius, R., Platen, E.: Time discrete Taylor approximations for Itô processes with jump component. Math. Nachr. **138**, 93–104 (1988)
25. Øksendal, B., Sulem, A.: Applied stochastic control of jump-diffusions. Universitext. Springer, Berlin (2005)
26. Platen, E.: An approximation method for a class of Itô processes with jump component. Liet. Mat. Rink. **22**(2), 124–136 (1982)
27. Platen, E., Bruti-Liberati, N.: Numerical Solution of Stochastic Differential Equations with Jumps in Finance. Springer, Berlin (2010)
28. Samuelides, Y., Nahum, E.: A tractable market model with jumps for pricing short-term interest rate derivatives. Quant. Finance **1**, 270–283 (2001)

Moving Least Squares for Arbitrage-Free Price and Volatility Surfaces

Pascal Heider

Abstract To price exotic options consistently to market data, it is necessary to approximate the implied volatility surface (IVS) over the strike–maturity plane. To avoid mis-pricing and arbitrage strategies, the approximation must be arbitrage free. Based on the moving least squares (MLS) reconstruction, a numerical approach is presented in this paper to compute arbitrage-free surfaces which approximate observed market data.

1 Introduction

Throughout the exposition, we assume that the underlying, which is most likely a stock index, follows a geometric Brownian motion with a constant dividend yield and a constant risk-free short rate. For any pair of strike K and maturity T of a stock option, there exists a unique model price which is denoted by $C(K, T)$. If not said otherwise, the value $C(K, T)$ is seen from today which is $t = 0$. Thus, this function C defines a surface in the strike–maturity plane, which is the *price surface*. Because there exists no continuum of traded options at the market, this surface is sampled at discrete points (K_i, T_i).

A typical application of the price surface is the consistent pricing of exotic options. Of course, not every construction is a meaningful solution and we have to guarantee that the constructed surface is arbitrage free, i.e., there are no trading strategies which generate risk-free financial gains. To fix ideas, let $p(S, T)$ denote the probability density of the underlying at time T with initial status S_0 at time $t = 0$

P. Heider (✉)

Universität zu Köln, Mathematisches Institut, 86–90, 50937 Köln, Weyertal, Germany

e-mail: pheider@math.uni-koeln.de

M. Cummins et al. (eds.), *Topics in Numerical Methods for Finance*, Springer Proceedings in Mathematics & Statistics 19, DOI 10.1007/978-1-4614-3433-7_2,

© Springer Science+Business Media New York 2012

under the risk-free measure. Then, the fundamental theorem of asset pricing yields
the explicit formula

$$C(K,T) = e^{-rT} \int_0^\infty (S_T - K)^+ p(S_T, T) dS_T \tag{1}$$

for the surface C.

From this representation, we can deduce several properties of the call surface,
[1,2]. Of course, the surface must be positive,

$$C(K,T) \geq 0 \tag{2}$$

and taking the derivatives with respect to K yields monotonicity and convexity of
the call-price in K,

$$-e^{-rT} \leq \frac{\partial C}{\partial K}(K,T) \leq 0 \tag{3}$$

$$\frac{\partial^2 C}{\partial K^2}(K,T) \geq 0. \tag{4}$$

Moreover, there is an arbitrage condition in maturity direction as well, [3,8],

$$\frac{\partial C}{\partial T} + (r-q)K\frac{\partial C}{\partial K} + qC(K,T) \geq 0. \tag{5}$$

We summarize these properties of $C(K,T)$ in a definition, [4]:

Definition 1. A positive-valued function $C(K,T)$ which is twice differentiable with
respect to the first variable and differentiable with respect to the second variable is
called *arbitrage free* with parameters r, q if it satisfies Eqs. (3)–(5).

The goal of the presented numerical method is to reconstruct an arbitrage-free
surface $C(K,T)$ out of a discrete set of observable market data. As observed by [5]
the implied volatility surface (IVS) is also arbitrage free if the call price surface is
arbitrage free from which the IVS was deduced. The IVS is obtained by inverting
the price surface with the Black–Scholes formula. The presented algorithm to find
an arbitrage-free surface is based on moving least squares (MLS) approximation,
[7], which is adapted to the above problem.

The remainder of the paper is organized as follows. First, we discuss in Sect. 2
the MLS approximation together with an extension to treat local constraints. The
application of the MLS approach to call price reconstruction is explained in detail
in Sect. 3. We show applications of the algorithm in Sect. 4 and compute price and
IVS for a given data set of observed market data. We summarize the results in Sect. 5
and finish with some final remarks.

2 Moving Least Squares and Local Constraints

Suppose that the values of a function $f : \Omega \to \mathbb{R}$ with $\Omega \subset \mathbb{R}^2$ are given at a discrete set of points $X = \{x_1, \ldots, x_N\} \subset \Omega$. An MLS approximation solves for every $x \in \Omega$ a weighted least squares fit. For that, let $\Phi_\delta : \Omega \to \mathbb{R}$ be a weight function, e.g., [6],

$$\Phi_\delta(x) = \exp\left(-\|S \cdot x\|_2^2/\delta^2\right), \qquad (6)$$

where δ is parameter, $\phi : [0, \infty) \to \mathbb{R}^+$ is a nonnegative function and $S \in \mathbb{R}^{2 \times 2}$ is a scaling matrix. A typical choice. For given $x \in \Omega$ the value $s_{f,X}(x)$ of the approximant is defined by $s_{f,X}(x) = p^*(x)$, where $p^*(x)$ is the solution of

$$\min_{p \in \Pi} \frac{1}{2} \sum_{j \in I(x)} (p(x_j) - f(x_j))^2 \cdot \Phi_\delta(x - x_j), \qquad (7)$$

where Π is a suitable polynomial space and

$$I(x) := \{i \in \{1, \ldots, N\} \mid \|S \cdot (x - x_i)\|_2 < \delta\}$$

contains the nearest neighbors of x in X. The next lemma summarizes a couple of properties of the MLS approximant, more details and proofs can be found in [9].

Lemma 1.

- *If $I(x)$ is Π_m uni-solvent, then the minimization is uniquely solvable and*

$$s_{f,X}(x) = p^*(x) = \sum_{i \in I(x)} a_i^*(x) f(x_i)$$

  for certain functions $a_i^*(x)$.
- *If $\Phi \in \mathscr{C}^k$, then $s_{f,X} \in \mathscr{C}^k$.*
- *If $f \in \mathscr{C}^{m+1}(\Omega^*)$ and X is sufficiently dense, then*

$$\|f - s_{f,X}\|_{L^\infty(\Omega)} \leq c \cdot h_{X,\Omega}^{m+1} |f|_{\mathscr{C}^{m+1}(\Omega)},$$

  where h is the fill distance of X which is the maximal distance for any x to the nearest data point from X.

The constraints are added to the equations locally. Hence, let $\mathscr{C}^{\text{fin}}$ be the function space of functions $f : \mathbb{R}^2 \to \mathbb{R}$ which are twice differentiable with respect to the first variable and once differentiable with respect to the second variable. Then, one can express l constraints on f by an operator $B : \mathscr{C}^{\text{fin}} \to \mathbb{R}^l$,

$$(Bf)(x) \leq 0 \quad \text{for all } x \in \Omega.$$

If we choose the polynomial space Π as a subspace of $\mathscr{C}^{fin}$, we can include these constraints into our MLS approximation by the augmented formulation

$$\min_{p \in \Pi} \frac{1}{2} \sum_{j \in I(x)} (p(x_j) - f(x_j))^2 \cdot \Phi_\delta(x - x_j)$$

$$\text{s. t.} \quad (Bp) \leq 0. \tag{8}$$

Thus, we are only considering such polynomials in the minimization process, which satisfy the constraints locally in x.

3 Computation of Arbitrage-Free Surfaces with MLS

We will apply the above procedure to approximate an arbitrage-free call price surface from observed market data. Let $C^{markt}(K_i, T_i) =: C_i$, $i = 1, \ldots, N$ denote the observed (scattered) call prices of calls with strike K_i and maturity T_i. Further, we assume that the short rate is constant for all times. Define the polynomial space $\Pi := \langle 1, \kappa, \kappa^2, \tau \rangle$ which is generated by the monomials $1, \kappa, \kappa^2, \tau$. This is the smallest polynomial space which satisfies $\Pi \subset \mathscr{C}^{fin}$. For fixed (K, T), one can represent any polynomial $c(\kappa, \tau) \in \Pi$ as

$$c(\kappa, \tau) = a_0 + a_1(\kappa - K) + a_2(\tau - T) + \frac{a_3}{2}(\kappa - K)^2 \tag{9}$$

with unique $\mathbf{a} = (a_0, a_1, a_2, a_3) \in \mathbb{R}^4$.

For fixed (K, T), define on Π the operator $\tilde{B}_{K,T} : \Pi \to \mathbb{R}^4$ by

$$\tilde{B}_{K,T}c = \begin{pmatrix} \frac{\partial c}{\partial \kappa}(K, T) \\ -\frac{\partial^2 c}{\partial \kappa^2}(K, T) \\ -\frac{\partial c}{\partial \tau} - (r - q)K\frac{\partial c}{\partial \kappa}(K, T) - qc(K, T) \\ -c(K, T) \end{pmatrix}, \tag{10}$$

which are the local constraints to guarantee arbitrage freeness by Definition 1. Moreover, define the index set $I(K, T) \subseteq \{1, \ldots, N\}$ by

$$I(K, T) := \left\{ i \in \{1, \ldots, N\} \,\middle\|\, \left\| S \cdot \begin{pmatrix} K_i - K \\ T_i - T \end{pmatrix} \right\|_2 < \delta \right\} \tag{11}$$

and define for every $j \in I(K, T)$ the differences between strike $\Delta_K^{(j)} := K_j - K$ resp. maturity $\Delta_T^{(j)} := T_j - T$ of the observed calls and the pair (K, T).

We formulate Eq. (8) as linearly constrained linear least square problem in the basis of the vector space Π and get, [4],

$$\min_{\mathbf{a}\in\mathbb{R}^4} \frac{1}{2}\left\|\Phi^{1/2}(A\mathbf{a}-b)\right\|_2^2$$

$$\text{s. t.}\quad B\mathbf{a}\leq 0 \tag{12}$$

with matrices $A\in\mathbb{R}^{|I(K,T)|\times 4}, B\in\mathbb{R}^{4\times 4}, \Phi\in\mathbb{R}^{|I(K,T)|\times|I(K,T)|}$ defined by

$$A:=\begin{pmatrix} 1 & \Delta_K^{(1)} & \Delta_T^{(1)} & \frac{1}{2}\left(\Delta_K^{(1)}\right)^2 \\ \vdots & & \vdots & \\ 1 & \Delta_K^{(|I(K,T)|)} & \Delta_T^{(|I(K,T)|)} & \frac{1}{2}\left(\Delta_K^{(|I(K,T)|)}\right)^2 \end{pmatrix}, \tag{13}$$

$$B_{K,T}\mathbf{a}=\begin{pmatrix} 0 & 1 & 0 & 0 \\ 0 & 0 & 0 & -1 \\ -q & -(r-q)K & -1 & 0 \\ -1 & 0 & 0 & 0 \end{pmatrix}\mathbf{a}, \tag{14}$$

$$\Phi:=\text{diag}\left(\phi\left(\left\|S\cdot\begin{pmatrix}K_j-K \\ T_j-T\end{pmatrix}\right\|_2\right)\middle| j\in I(K,T)\right), \tag{15}$$

and the data vector

$$b:=(C_j)_{j\in I(K,T)}\in\mathbb{R}^{|I(K,T)|}. \tag{16}$$

If the points $(K_j,T_j)\in\mathbb{R}^2$, $j\in I(K,T)$ are Π-unisolvent, i.e., the only polynomial $p\in\Pi$ which vanishes in all points (K_j,T_j) is the zero polynomial, then minimization problem (12) has a unique solution $\mathbf{a}^*$, [4]. Consequently, there exists a unique polynomial $c^*\in\Pi$ which solves minimization problem (8). Hence, we can define point-wisely the function

$$C(K,T):=c^*(K,T)=a_0^*, \tag{17}$$

for every pair (K,T) and this surface C is arbitrage free, [4]. Numerically, surface $C(K,T)$ can be evaluated easily and efficiently. The index set $I(K,T)$ can be computed by linear search in $O(N)$ time. The minimization problem (12) can be solved by any standard solver for linear least squares problems.

4 Numerical Study

Our data set consists of call options on the DAX written by Deutsche Bank. The data were obtained via the online portal db-x market (http://www.x-markets.db.com/) and contain the bid and ask prices from July, 29th, 2009 of 1,304 options with different strikes and maturities, see Fig. 1a. The risk-free short rate was set constant to the

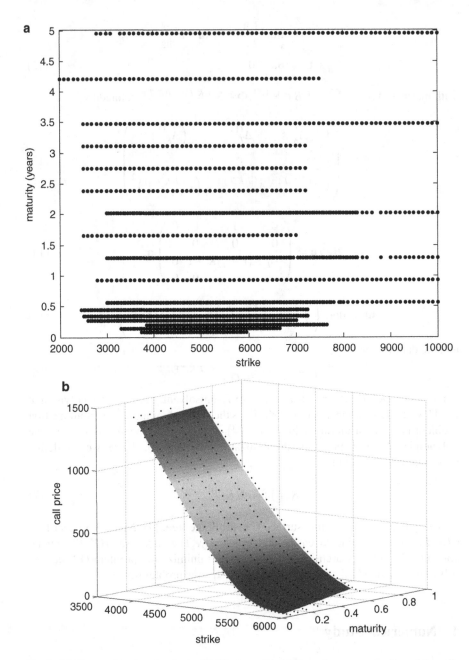

Fig. 1 (a) Observed market data in the (K, T)-plane. (b) Arbitrage-free approximation of the call surface together with some data points

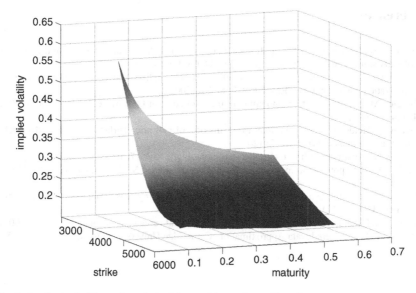

Fig. 2 Implied volatility surface derived from arbitrage-free call surface

Euribor $r = 0.4\%$ of that day. The index was at $S_0 = 5,268.51$ points and we assume 250 trading days per year. The smooth arbitrage-free call surface to this data set is shown in Fig. 1b together with some data points.

The IVS can be computed from the call surface by inverting the Black–Scholes equation with respect to the volatility. The IVS shown in Fig. 2 is deduced from the previous example by inverting the Black–Scholes formula.

5 Conclusion

We have presented a robust and flexible numerical method to compute provable arbitrage-free call price surfaces from observed market data. The method is based on a local polynomial reproduction with local constraints to enforce the no-arbitrage conditions. The approximating polynomial space is chosen large enough to capture relevant financial behavior and small enough to ensure stability of the algorithm. The algorithm computes a local representation of the call price surface in the approximating polynomial space so that the derivatives of the surface are also computed as a by-product. From these data, one can directly obtain the IVS, the local volatility surface, and the transition probabilities.

References

1. Brunner, B., Hafner, R.: Arbitrage-free estimation of the risk-neutral density from the implied volatility smile. J. Comput. Finance **7**(1), 75–106 (2003)
2. Bertsimas, D., Popescu, I.: On the relation between option and stock prices: a convex optimization approach. Oper. Res. **50**(2), 358–374 (2000)
3. Fengler, M.R.: Arbitrage-free smoothing of the implied volatility surface. Quant. Finance **9**(4), 417–428 (2009)
4. Glaser, J., Heider, P.: Arbitrage-free approximation of call price surfaces and input data risk. Quantitative Finance, **12**(1), 61–73 (2012)
5. Kahalé, N.: An arbitrage-free interpolation of volatilities. Risk **17**, 102–106 (2004)
6. Levin, D.: The approximation power of moving least-squares. Math. Comput. **67**(224), 1517–1531 (1998)
7. Lancaster, P., Salkauskas, K.: Surfaces generated by moving least squares methods. Math. Comput. **37**(155), 141–158 (1981)
8. Reiner, E.: Calendar spreads, characteristic functions, and variance interpolation. Mimeo (2000)
9. Wendland, H.: Scattered Data Approximation. Cambridge University Press, Cambridge (2005)

Solving Impulse-Control Problems
with Control Delays

Kumar Muthuraman and Qi Wu

Abstract Several classes of stochastic control problems, wherein the controller can bring about an instantaneous change in state, are called free boundary problems. Moving boundary methods are a class of computational methods that have been developed recently to solve such free boundary problems. The goal of this paper is to provide a detailed description of the methodology. We specifically focus on stochastic impulse-control problems which arise when the cost of control includes a fixed cost. The inclusion of a fixed cost, very common in financial applications, makes the control effect finite changes in state, bringing about discontinuities in the state evolution. These problems are, hence, more complicated than problems wherein controls have only proportional costs (singular control) or wherein controls simply terminate the process (optimal stopping). We show how the impulse-control problem is transformed to a Quasi Variational Inequality and then describe the moving boundary method. We demonstrate problems with no control delay, fixed delay and stochastic delay. We also review all the theoretical guarantees that have been established. This paper summarizes and presents an implementation focused description of the research presented in (Feng and Muthuraman, A computational method for stochastic impulse control problems. Mathematics of Operations Research **35**(4), 830–850, 2010 and Muthuraman et al., Inventory management with stochastic lead times. Working Paper, 2011) for solving impulse-control problems.

K. Muthuraman (✉) • Q. Wu
McCombs School of Business, University of Texas, Austin, TX 78712
e-mail: kumar.muthuraman@mccombs.utexas.edu; qi.wu@phd.mccombs.utexas.edus

M. Cummins et al. (eds.), *Topics in Numerical Methods for Finance*, Springer Proceedings
in Mathematics & Statistics 19, DOI 10.1007/978-1-4614-3433-7_3,

1 Introduction

Stochastic control problems are those wherein the controller seeks control an evolving stochastic process in order to maximize (or minimize) an objective. Stochastic control problems are usually transformed to differential equations by using dynamic programming arguments. When the controller has the ability to instantaneously displace the state, rather than make changes only to the rate of change, the differential equations resulting from the transformation usually take the form of a free-boundary problem. Free-boundary problems are those wherein one or more of the boundaries of the domain are unknown and need to be computed as a part of the solution.

When the cost of control has a fixed component associated with it, it is usually optimal for the controller to bring about non-infinitesimal changes to the state and, hence, state evolutions would become discontinuous. Such problems wherein the optimal control brings about jumps in state are called impulse-control problems as opposed to singular control problems wherein the optimal control only makes infinitesimal changes in state. Singular control problems usually have only proportional control costs. Optimal stopping problems are another class of stochastic control problems where the only control is to terminate the stochastic process.

Moving boundary methods where first developed for singular control problems [6, 8] and have also been shown to be very efficient for optimal stopping problem [7, 11]. The approach has also been adequately surveyed in [9], for both singular and optimal control. In this paper, we summarize and detail the methodology for three canonical impulse-control formulations of increasing complexity. In the first formulation, we do not include a control lags. In the second formulation, we allow for a fixed lag and the third allows for a stochastic lag. In each of these cases, we show how one can use the moving boundary method to solve the problem. This paper essentially presents an implementation focused impulse-control solution procedure for various kinds of impulse-control problems by summarizing in one place the relevant parts of the research presented in [5, 10].

1.1 Structure and Intended Audience

This paper briefly summarizes the formulation and the primary theoretical guarantees that have been established but does not elaborate on these. The focus is on the nuts and bolts of solving the free-boundary differential equations that characterizes the optimal impulse control. To this extent, we have also provided a MATLAB code that implements the moving boundary procedure described in this paper.

In the rest of this section, we provide a brief literature review. Section 2 describes the formulation, transformation to the differential equations problem (often called the Quasi-variational-inequality, QVI), and the moving boundary algorithm to solve the QVI. While Sect. 2 restricts attention to the case with no control delays, Sect. 3 focuses on adding fixed and stochastic delays.

1.2 Related Literature

We briefly summarize in this section some applications of impulse-control problems in literature. Impulse-control problems are often interpreted as inventory-control formulations with the stochastic process describing the evolution of warehouse inventory. Sulem [13] considers an inventory-control problem where demand follows a Wiener Process. The ongoing inventory-holding cost includes the storage and shortage cost. Is it shown that the (s, S) policy, where the controller increases the inventory to S whenever inventory falls to s, is optimal. Later in [1], they extend the previous paper to an inventory management problem with a fixed lag in deliveries. By changing the state variable from the inventory level to inventory position and using a sunk cost argument, they were able to show that the inventory problem with fixed lead time can be converted to the inventory-control problem with no lead time, with the holding cost function replaced by a convolution with the lead time demand.

Another popular class of impulse-control problems are described as the cash management problems where the state is interpreted as the evolution of the cash on hand of a firm. There are fixed and proportional costs for adjusting the liquid cash and the controller seeks to maintain an optimal level. Bar-Ilana et al. [2] and Constantinides and Richard [4] study the cash management problem. There are several other applications. For example, [3] use real options base impulse-control models to determine a firm's investment decision on the natural resources. They assumed that besides production level, the firm can make decision on suspending the production and closing production fields temporarily, or abandoning completely. Since there are fixed costs associated with closing or reopening production, this results in an impulse-control problem.

2 Model Formulation of Impulse-Control Problems

Say the uncontrolled stochastic process $\widetilde{Y}(t)$ that starts at x has dynamics that can be represented as $d\widetilde{Y}(t) = \mu dt + \sigma dW_t$. Here W_t is a Wiener process in $\mathbf{R}$ and $\mathscr{F}_t$ be the increasing family of σ-algebra generated by W_t. Let $0 \leq \tau_1 \leq \tau_2 \leq \cdots \leq \tau_i \leq \cdots$ be a sequence of stopping times adapted to $\{\mathscr{F}_t\}$, such that only a finite number of τ_i will occur in a bounded interval. An admissible impulse control v is defined as:

$$v = (\tau_1, \xi_1; \cdots ; \tau_i, \xi_i; \cdots), \tag{1}$$

where ξ_i's are the instantaneous change to the states at time τ_i. With the control v, the dynamic of the control process, Y, is given by

$$dY(t) = \mu dt + \sigma dW_t + \xi_i \mathbf{1}_{t=\tau_i}$$

with $Y|_{0-} = x$.

The control costs have two parts. One is the fixed cost, which is independent of the control magnitude $|\xi|$, but depends on whether we are increasing or decreasing state variable. The other cost is the proportional control cost, which depends not only on if the state decreases or increases, but also on the control magnitude. Besides the control cost, the system incurs an instantaneous carrying cost, which depends on the states of the controlled process $Y(t)$.

Assuming the discount rate is α, the discounted total cost of the system is:

$$F_v(x) = E\left\{ \int_0^\infty h(Y(t))e^{-\alpha t}dt + \sum_n c(\xi_n)e^{-\alpha \tau_n} \right\}, \tag{2}$$

where $c(\xi)$ is the control cost, and can be written as

$$c(\xi) = \begin{cases} K + k \cdot \xi & \xi > 0 \\ L + l \cdot |\xi| & \xi < 0 \\ 0 & \text{otherwise.} \end{cases} \tag{3}$$

Our objective is to choose an optimal control policy v to minimize the total cost $F_v(x)$.

2.1 QVI, Verification Theorem

Let $V(x)$ denote the minimum cost under the optimal policy, i.e., $V(x) = \inf_v F_v(x)$. The QVI that characterizes $V(x)$ can be derived by using standard dynamic programming arguments that goes as follows. If we adopt an arbitrary control for an infinitesimal amount of time and then switch back to the optimal control, then the resulting value function cannot be better than the optimal one. This provides us a relation that characterizes $V(x)$. Because of the fixed control cost and the nature of impulse control, we only have three choices of control during that infinitesimal period: not impose any control, impose an impulse control to increase state, or impose an impulse control to decrease state.

Under the first choice, we have:

$$V(x) \leq \int_0^{\Delta t} h(x_t)dt + E[V(x_{\Delta t})e^{-\alpha \Delta t}]. \tag{4}$$

Next, say we increase state by ξ units and get back to the optimal control thereafter. We then have:

$$V(x) \leq K + k\xi + V(x+\xi). \tag{5}$$

Since the above inequality holds for any instantaneous control ξ, we have

$$V(x) \leq \inf_{\xi}\left(K + k\xi + V(x+\xi)\right). \tag{6}$$

Similarly, say we decrease state by $|\xi_i|$ units, then, we have:

$$V(x) \leq \inf_{\xi}\left(L + l|\xi| + V(x-\xi)\right). \tag{7}$$

One of the three inequalities (4)–(7) must hold, since one of these three choices of control must be optimal.

Using the above argument and applying Itô's formula to (4), we have:

$$\Gamma V(x) = \frac{1}{2}\sigma^2 V''(x) + \mu \cdot V'(x) - \beta \cdot V(x) \geq -h(x),$$

$$\mathcal{M}V(x) = \inf_{\xi>0}\{V(x+\xi) + K + k \cdot (\xi)\} - V(x) \geq 0,$$

$$mV(x) = \inf_{\xi>0}\{V(x-\xi) + L + l \cdot |\xi|\} - V(x) \geq 0.$$

For ease of notation, let $\mathcal{L}V(x) = \Gamma V(x) + h(x)$. For the Verification Theorem, we refer readers to [12] for the following result:

Theorem 1. *Suppose $f(x) \geq 0$, $f'(x)$ is absolutely continuous and bounded, $f''(x)$ exists, a.e., $f''(x) \in L^2(\mathbf{R})$. If f satisfies*

$$\mathcal{L}f(x) \geq 0 \ a.e. \ x, \tag{8}$$

$$\mathcal{M}f(x) \geq 0, \tag{9}$$

$$mf(x) \geq 0, \tag{10}$$

and for any given $x \in R$, at least one of the above inequalities becomes equality, then,

$$f(x) = \mathcal{F}_x(\hat{v}) \leq \mathcal{F}_x(v), \quad \forall \ admissible \ v$$

is the value function. The corresponding optimal control $\hat{v}$ is then defined by $\tau_1 = \inf_{t \geq 0}\{\tilde{Y}(t) \notin \mathcal{C}\}$ and

$$\xi_1 = \begin{cases} arginf_{\eta>0}\{f(Y(\tau_1-)+\eta)+K+k\cdot\eta\} & if \ \mathcal{M}f(Y(\tau_1-)) = 0, \\ arginf_{\eta>0}\{f(Y(\tau_1-)-\eta)+L+l\cdot\eta\} & if \ otherwise. \end{cases}$$

The theorem not only provides us the sufficient condition for optimality, but also describes the optimal policy. The optimal control is to keep the state space in $\mathscr{C}$. If the control process is in the interior of the region, then we do nothing. The moment the process attempts to leave $\mathscr{C}$, the control immediately exerts a significant change in state.

2.2 Free Boundary Problems and (d,D,U,u) Policy

Consider a class of control policies that can be characterized using four parameters: d, D, U, and u. When the state variable goes down and strikes the lower bound value d, the control lifts the state up to D. When state increases to strike the upper bound u, the control brings the state variable down to U. The controlled process, hence, is always kept in (d, u).

The following theorem [5] guarantees that under conditions listed, the optimal policy belongs to the class of (d, D, U, u) policies. Therefore, it is sufficient to consider controls in this class and search for the optimal parameters d, D, U, and u.

Theorem 2. *Assume that*

  (i) *$h(x)$ is convex, $C(\mathbf{R}) \cap C^1((-\infty, 0) \cup (0, +\infty))$ with the minimum achieved at $x = 0$.*
 (ii) *When x is small enough, $h'(x) < -\beta \cdot k$; when x is large enough, $h'(x) > \beta \cdot l$.*
(iii) *There exists $d < D < U < u$, and V is the cost function associated with the policy v characterized by (d, D, U, u).*
 (iv) *$V'(x)$ is continuous (and thus $V'(d+) = V'(d-) = -k$ and $V'(u-) = V'(u+) = l$).*
  (v) *$\exists \varepsilon_1, \varepsilon_2 > 0$, such that $V'(x) + k < 0$, $\forall x \in (d, d + \varepsilon_1]$, and $-V'(y) + l < 0$, $\forall y \in [u - \varepsilon_2, u)$.*
 (vi) *$D = argmin_{x \in (d,u)}\{V(x) + k \cdot x\}$, which implies $V'(D) = -k$.*
(vii) *$U = argmin_{x \in (d,u)}\{V(x) - l \cdot x\}$, which implies $V'(U) = l$.*

Then V is the value function, and (d, D, U, u) is the corresponding optimal policy.

Feng and Muthuraman [5] also show that for any given admissible policy (d, D, U, u), the corresponding cost function $v(x)$ satisfies

$$\mathscr{L}v(x) = 0, \quad \forall x \in (d, u), \tag{11}$$

$$v(x) = v(D) + K + k \cdot (D - x), \quad \forall x \leq d, \tag{12}$$

$$v(x) = v(U) + L + l \cdot (x - U), \quad \forall x \geq u. \tag{13}$$

That is, for a given policy (d, D, U, u), the associated cost function can be obtained by solving a fixed boundary differential equation. Next we show this algorithm results in policy improvement. Due to the above two theorems, each step of the following algorithm is well defined and converges to the optimal value functions.

2.3 Algorithm

First, we start with an initial guess (d_0, D_0, U_0, u_0), and solve for the expected cost V_0 under the initial policy (d_0, D_0, U_0, u_0). For the initial guess, we need to make sure that $V_0'(d_0) + k \geq 0$ to guarantee $d_0 < d^*$, where d^* is the d in the optimal policy. Similarly, we need to make sure $-V_0'(u_0) + l \geq 0$ to guarantee $u_0 > u^*$, where u^* is the u in the optimal policy, since the domain is fixed as (d_0, u_0). The solution to the fixed boundary ordinary differential equations are computed using the finite difference method. Using V_0, we update the boundaries by $d_1 = \sup\{d \in [d_0, D_0) : \forall x \in [d_0, d], V_n'(x) + k \geq 0\}$, $u_1 = \inf\{u \in (U_0, u_0] : \forall x \in [u, u_0], -V_n'(x) + l \geq 0\}$. After obtaining the new boundaries d_1 and u_1, we solve the fixed boundary ODE again with the policy (d_1, D_0, U_0, u_1) to get updated cost function $\bar{V}^0$. Then, we update D by $D_1 = \mathrm{argmin}_{u_1 > x > d_1}\{\bar{V}_n(x) + k \cdot x\}$, and U by $U_1 = \mathrm{argmin}_{d_1 < x < u_1}\{\bar{V}_n(x) - l \cdot x\}$. In two updates, we obtain the updated inventory policy (d_1, D_1, U_1, u_1).

We are guaranteed that (d_1, D_1, U_1, u_1) is an improved policy compared to the initial guess. Moreover, $d_1 < d^*$ and $u_1 > u^*$ too, and therefore we can iterate the above process until it converges. Feng and Muthuraman [5] prove that this iterative process converges and the cost function is in C^1, and that the converged policy is the solution to the free boundary problem.

2.4 Inventory-Control Application

In this section, we will take examples in inventory-control problems with stochastic demand and continuous review. Here, we will consider the case with no delivery lag and in Sect. 3 consider the cases with fixed delivery lag and stochastic delivery lag.

We model the stochastic demand using the simple Itô process:

$$dD_t = \mu dt + \sigma dW_t. \tag{14}$$

Then the cumulative demand D_t follows a Normal distribution $N(\mu t, \sigma^2 t)$. Inventory position x_t can then be written as:

$$dx_t = -\mu dt - \sigma dW_t + dv_t, \tag{15}$$

where v_t is the cumulative order arrival process demand as $dv_t = \xi_t$, if $t \in \{\eta_i\}_{i \in \mathcal{N}}$; $dv_t = 0$, otherwise.

Replenishing the inventory for ξ units results in a fixed cost K and a variable cost $k\xi$. Since generally discarding current inventory is not considered as an admissible control, the inventory-control problems only have one side control. In other words, the control of replenishment can only bring up the inventory level. It is a special

Table 1 Inventory
management example

n	s_n	S_n	$V'_n(s_n)+k$
1	−5.000	0.000	1.599
2	−2.574	1.872	0.954
3	−1.102	0.925	0.338
4	−0.577	0.565	0.088
5	−0.441	0.493	0.010
6	−0.431	0.489	0.003
7	−0.426	0.489	0.000

case of the (d,D,U,u) policy, with $s=d$, $S=D$, and $U=u=\infty$. The carrying cost is in the form of unit storage cost q, when the inventory level is positive, while it is in the form of unit shortage cost p, when the inventory level is negative.

The total cost function is

$$E\left\{\sum_{i=0}^{\infty}c(\xi_i)e^{-\alpha\eta_i}+\int_0^{\infty}e^{-\alpha s}h\left(x+\sum_{\eta_i\in(0,s)}\xi_i+D_{(0,s]}\right)\mathrm{d}s\right\}. \qquad (16)$$

For the immediate delivery, from the QVI formulation derived above, the value function V solves:

$$\max[\mathscr{L}V,\mathscr{M}V]=0, \qquad (17)$$

where $\mathscr{L}V=-\frac{1}{2}\sigma^2\frac{\partial^2 V}{\partial x}+\mu\frac{\partial V}{\partial x}+\alpha V-h(x)$, $\mathscr{M}V=V(x)-\left(K+\inf_{\xi>0}[k\xi+V(x+\xi)]\right)$.

Using the algorithm in Sect. 2.3, it is straightforward to solve (17). The code is provided in the appendix. Consider the parameter set with discount factor $\alpha=0.01$, average demand rate $\mu=-0.3$, demand volatility $\sigma=0.2$, shortage cost $p=0.2$, storage cost $q=0.11$, variable ordering cost $k=1$, fixed ordering cost $K=0.1$, discretization parameter $\mathrm{d}x=0.1$ by setting M $=10$, we set the domain set to be $[-10,10]$. Using these parameters and calling the function "inventory.m" directly, we get the optimal order policy $(-0.426,0.489)$, i.e., to bring the inventory level to 0.489, once the inventory drops to -0.426. The values of (s_n,S_n) and the convergent criterion $V'_n(s_n)+k$ are listed in Table 1 for all iterations.

3 Impulse Control with Fixed Delay and Stochastic Delay

Immediate delivery in inventory control is usually a simplified and idealistic assumption. In reality, there are processing times and transportation times that lead to the delivery lag in inventory replenishment. We show in this section that after changing the state variable from the inventory level to the inventory position, the optimal control problem of fixed lead times is in the exact same format as the one in the immediate delivery case.

From earlier development, when there is fixed lead time τ, the optimal value function can be written as

$$V(x) = \inf_{v \in \mathcal{V}_{ad}} E \left\{ \sum_{i=0}^{\infty} c(\xi_i) e^{-\beta \eta_i} + \int_{\tau}^{\infty} e^{-\beta s} h \left(x + \sum_{\eta_i \in (0,s-\tau]} \xi_i - D_{(0,s]} \right) ds \right\}.$$

The expectation in the second term is taken over $D_{(0,s]}$ for $s \geq \tau_0$. Writing the expectation with respect to $D_{(0,s]}$ as an integral over the demand with probability distribution $F(D_{(0,s]})$, one can write the second term as:

$$\int_{\tau}^{\infty} \int e^{-\beta s} h \left(x + \sum_{\eta_i \in (0,s-\tau]} \xi_i - D_{(0,s]} \right) dF_{D_{(0,s]}} ds$$

$$= \int_{\tau}^{\infty} e^{-\beta s} \int \left[\int h \left(x + \sum_{\eta_i \in (0,s-\tau]} \xi_i - D_{(0,\tau]} - D_{(\tau,s]} \right) dF_{D_{(0,\tau]}} \right] dF_{D_{(\tau,s]}} ds. \quad (18)$$

Define the new holding cost function: $\tilde{h}(z) = e^{-\beta \tau} \int h(z - D_{(0,\tau]}) dF_{D_{(0,\tau]}}$, and since D is a time-invariant process, the above expression can be written as:

$$E \left\{ \int \int \int_{\tau}^{\infty} e^{-\beta(s-\tau)} \tilde{h} \left(x + \sum_{\eta_i \in (0,s-\tau]} \xi_i - D_{(0,s-\tau]} \right) ds \cdot dF_{D_{(0,s-\tau]}} \right\}. \quad (19)$$

Define new variables $\tilde{s} = s - \tau$. Then, the above expression can be written as:

$$E \left\{ \int_{0}^{\infty} e^{-\beta \tilde{s}} \tilde{h} \left(x + \sum_{\eta_i \in (0,\tilde{s})} \xi_i - D_{(0,\tilde{s}]} \right) d\tilde{s} \right\}, \quad (20)$$

where the expectation is taken over $D_{[0,\tilde{s})}$. If we add back the ordering cost, then the value function would be in a form identical to (16). Therefore, the QVI formulation of this problem is exactly the same as (17) except that the $h(x)$ in $\mathcal{L}V$ definition is no longer a simple inventory-holding cost $h(x)$. Instead, it becomes $\tilde{h}(z) = e^{-\beta \tau} \int h(z - D_{(0,\tau]}) dF_{D_{(0,\tau]}}$. The code in appendix still applies to this case, except the right-hand side need to be adapted to the definition of $\tilde{h}$.

While after the transformation the inventory control with the fixed lead time is relatively straightforward to solve, inventory control with stochastic lead time is much more challenging. Because of uncertain order arrival time of outstanding orders, the history information about the order placement time and the arrival history of past orders become relevant. This results in a severe dimensionality problem.

Muthuraman et al. [10] shows that when the lead time satisfies a set of mild assumptions, the dimension of such a system's state space can be reduced to one, i.e., the system only depends on the inventory position. With some additional assumptions, they are able to show that the order delivery process is a pure jump process with Poisson jump arrivals. Based on this paper, we show in this section how our numerical method can be applied in the stochastic control lead time cases.

Under a stochastic lead time, the retailer cannot predict when the order arrives. Let τ_t denote the arrival time for an order that might have been placed at time t, then $\tau_t - t$ is defined for all t and therefore referred to as the virtual lead time [15]. They assume that τ_t is a process that is left continuous with right limit almost surely. Moreover, they define $w_t = \tau_t - t$ as the waiting time for the order placed at time t.

To further capture the nature of stochastic order arrival, define B_t as the youngest virtual order, all the orders before which are fulfilled at time t. It can be expressed by τ_t in the way that $B_t = \inf\{s | \tau_s > t\}$, while τ_t can be expressed $\tau_t = \sup\{s | B_s < t\}$. Let $\mathscr{F}_t$ denote the filtration of the stochastic process B up to time t, which contains all the order arrival history on $(-\infty, t]$, i.e., $\mathscr{F}_t = \sigma\{B_s, s \leq t\}$. While B_t is a backward tracking variable, τ_t and w_t are forward looking. Let us call B_t the order delivery process and τ_t the order arrival process. Zipkin [14] uses similar variables to model the stochastic lead time.

In this case, the cost function can be written as:

$$E\left\{\sum_{i=0}^{\infty} c(\xi_i)e^{-\alpha\eta_i} + \int_{\tau_0}^{\infty} e^{-\alpha s} h\left(x + \sum_{\eta_i \in (0, B_s]} \xi_i - D_{(0,s]}\right) ds\right\}.$$

They show that, under some mild assumptions, the dimensionality can be reduced and the optimal inventory policy only depends on the inventory position x_t. Those assumptions assume that there is no dependency of the inventory position on the time, ordering history and order arrival history. Moreover, they also show that with additional assumptions, the order delivery process is the a pure jump process with poisson jump arrivals.

The central idea in the analysis is to separate total cost into the sunk cost and effective cost. Since any cost the system incurs before τ_0 cannot be affected by current or further controls, this part of the cost is sunk cost. The cost starts from τ_0 is the effective cost. Let $V(x)$ be the optimal effective cost. Then applying dynamic programming arguments, we obtain the QVI

$$V(x_0) \leq E\left[\int_{\tau_0}^{\tau_{\Delta t}} e^{-\alpha\tilde{\tau}} h(x_0 - D_{[0,\tilde{\tau})}) d\tilde{\tau}\right] + E[V(x_{\Delta t})e^{-\alpha(\tau_{\Delta t} - \tau_0)}]$$

$$\leq \int_0^{\infty} \int_{\tau_0}^{\infty} \int_{-\infty}^{\infty} \int_{\tau_0}^{\tau_{\Delta t}} e^{-\alpha\tilde{\tau}} h(x_0 - D)\phi_{\tilde{\tau}}(D) d\tilde{\tau} \cdot dD \cdot f(\tau_{\Delta t}, \tau_0) d\tau_{\Delta t} d\tau_0$$

$$+ E[V(x_{\Delta t})e^{-\alpha(\tau_{\Delta t} - \tau_0)}], \tag{21}$$

$$V(x_0) \leq K + \inf_{\xi > 0}(k\xi + V(x_0 + \xi)). \tag{22}$$

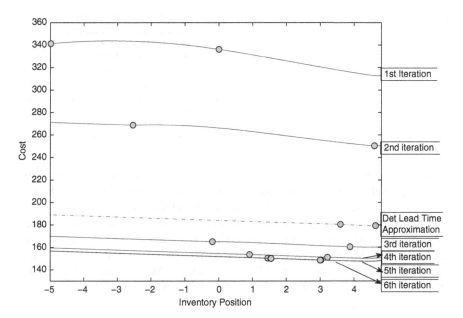

Fig. 1 Cost function for each policy

One of the two inequalities (21), (22) should be tight, since placing an order or not are the only two choices for the decision maker. We can simplify (21) and finally express the simplified version of (21) and (22) in a QVI form same as in (17) again. However, in this case

$$\mathscr{L}V = -\frac{1}{2}\sigma^2\frac{\mathrm{d}^2V}{\mathrm{d}x^2} + \mu\frac{\mathrm{d}V}{\mathrm{d}x} + \alpha V - H(x),$$

$$MV = V(x) - \left(K + \inf_{\xi>0}[k\xi + V(x+\xi)]\right),$$

where

$$H(x) = \int_0^\infty \left(\int_{\tau_0}^\infty \int_{\tau_0}^{\tau_{0+}} \int_{-\infty}^\infty e^{-\alpha\tilde{\tau}} h(x-D)\phi_{\tilde{\tau}}(D)\mathrm{d}D \cdot \mathrm{d}\tilde{\tau} \cdot f(\tau_{0+}|\tau_{0+} \neq \tau_0, \tau_0)\mathrm{d}\tau_{0+}\right)$$

$$\times \mu(\tau_0)f(\tau_0)\mathrm{d}\tau_0.$$

Therefore, stochastic lead time in the impulse-control problem only has an effect on the right-hand side of the differential equation. The major computational step is exactly the same as in the no delay case. Therefore, for the stochastic control lag case, we only need to adapt the right-hand side of the differential equation (*temp*) to the definition of H.

In Fig. 1, we draw the cost function corresponding to the inventory strategies we get in each iteration. The two points we highlight on the cost functions are

Table 2 Sequence of
policies

n	s_{stoc}	S_{stoc}	s_{det}	S_{det}
0	−5.00	0.00	−5.0	0.00
1	−2.54	4.60	−0.5	6.80
2	−0.19	3.88	2.00	5.48
3	0.91	3.21	3.48	4.66
4	1.45	3.02	3.59	4.63
5	1.54	3.00	3.59	4.63
6	1.55	3.00	3.59	4.63

values corresponding to s_{stoc} and S_{stoc}) (Table 2). The cost function when using the suboptimal strategy that assumes the deterministic lead time (set to be the mean of stochastic lead time) is also computed and shown in Fig. 1. We denote the optimal average inventory cost as V_{stoc}, and the average inventory cost under the approximated inventory strategy to be V_{det}. We use $\Delta V = (V_{\text{det}} - V_{\text{stoc}})/V_{\text{stoc}}$ as our measure of the relative difference between the optimal cost and the cost of the suboptimal policy $(s_{\text{det}}, S_{\text{det}})$. From Fig. 1, we can see ΔV is around 18 %, i.e., the approximated inventory strategy increases the inventory cost by 18 %.

4   Concluding Remarks

The computational methodology described in this paper is far more general and can be adapted for a variety of stochastic control problems that can be characterized by ordinary or partial differential equations. The adaptation requires some understanding and tweaking of the update procedure, which makes the method somewhat problem specific. However, this also allows the method to leverage on the problem structure to improve tremendously on computational efficiency. While this paper describes the methodology using some impulse-control examples, [9] contains detailed descriptions of the method for singular and optimal stopping examples.

Appendix

```
function [sV,SV,meanV,sVValue, nsimend]=inventory(mu, sigma,dx,
    M, K,k, p,q, alpha)
% example: inventory (0.4, 0.3, 0.01, 20, 0.1, 1, 0.1, 0.7,
    %0.5, 0.01)
%% initial value
S=0;
s=-M+0.01;
Nsim=100;
sV=zeros(Nsim,1);
SV=zeros(Nsim,1);
```

```
sV(1)=s;
SV(1)=S;

D=-M:dx:M;
temp=D.*(D>0)*q-D.*(D<0)*p;
predx= log10(dx); pretemp=0.1;
for nsim=2:Nsim
    b=-temp( int32( (s-sV(1))/dx+1):end );
    N=length(b);
    A1=[1 zeros(1,N-1)];
    A1(1,int32((S-s)/dx+1))=-1;
    b(1)= K+k*(S-s);
    A2last= sparse( (1:(N-2)), (1:(N-2)),(0.5*sigma^2/dx^2) );
    A2first= sparse( (1:(N-2)), (1:(N-2)),
        (0.5*sigma^2/dx^2+mu/dx));
    A2mid = sparse( (1:(N-2)), (1:(N-2)),
        (-sigma^2/dx^2-alpha-mu/dx) );
    A2=[A2first zeros(N-2,2)]
        +[zeros(N-2,1)  A2mid, zeros(N-2,1)]+...
        [zeros(N-2,2) A2last] ;
    A3=[zeros(1,N-2) -1 1];
    b(end)=q/alpha*dx;
    A=[A1;A2;A3];
    gnew=A\b;
    g = [(gnew(1) + k* (roundn((sV(nsim-1)-sV(1))
        /dx):-1:1)*dx)';
gnew];

    if mod(nsim,2)==0
        dgnew=(gnew(2:end)-gnew(1:(length(gnew)-1)))/dx;
        for (i=1:((S-s)/dx))
            if dgnew(i)<-k;
                break;
            end
        end
        s = roundn(s+(i-1)*dx, predx);
        sV(nsim)=s;
        SV(nsim)=S;
        dsV(nsim) = dgnew(i)+k;
    else
        [ming,minindex]=min(gnew+k*(s:dx:M)');
        S = roundn(s+(minindex-1)*dx, predx);
        SV(nsim)=S;
        sV(nsim)=s;
    end
     if(pretemp+abs(sV(nsim)-sV(nsim-1))+abs(SV(nsim)
        -SV(nsim-1))<dx/2);
         break;
     end
    pretemp=abs(sV(nsim)-sV(nsim-1))+abs(SV(nsim)-SV(nsim-1));
end
```

References

1. Bar-Ilan, A., Sulem, A.: Explicit solution of inventory problems with delivery lags. Math. Oper. Res. **20**(3), 709–720 (1995)
2. Bar-Ilana, A., Perryb, D., Stadjec, W.: A generalized impulse control model of cash management. J. Econ. Dynam. Contr. **28**, 1113–1133 (2004)
3. Brennan, M.J., Schwartz, E.S.: Evaluating natural resource investments. J. Business **58**(2), 135–157 (1985)
4. Constantinides, G.M., Richard, S.F.: Existence of optimal simple policies for discounted-cost inventory and cash management in continuous time. Oper. Res. **26**(4), 620–636 (1978)
5. Feng, H., Muthuraman, K.: A computational method for stochastic impulse control problems. Math. Oper. Res. **35**(4), 830–850 (2010)
6. Kumar, S., Muthuraman, K.: A numerical method for solving stochastic singular control problems. Oper. Res. **52**(4), 563–582 (2004)
7. Muthuraman, K.: A moving boundary approach to American option pricing. J. Econ. Dynam. Contr. **32**(11), 3520–3537 (2008)
8. Muthuraman, K., Kumar, S.: Multi-dimensional portfolio optimization with proportional transaction costs. Math. Finance **16**(2), 301–335 (2006)
9. Muthuraman, K., Kumar, S.: Solving free-boundary problems with application in finance. Foundations and Trends in Stochastic Systems **1**(4), 259–341 (2008)
10. Muthuraman, K., Seshadri, S., Wu, Q.: Inventory management with stochastic lead times. Working Paper (2011)
11. Muthuraman, K., Zha, H.: Simulation based portfolio optimization for large portfolios with transaction costs. Math. Finance **18**(1), 115–134 (2008)
12. Richard, S.F.: Optimal impulse control of a diffusion process with both fixed and proportional costs of control. SIAM J. Contr. Optim. **15**, 77–91 (1977)
13. Sulem, A.: A solvable one-dimensional model of a diffusion inventory system. Math. Oper. Res. **11**(1), 125–133 (1986)
14. Zipkin, P.H.: Stochastic lead times in continuous-time inventory models. Nav. Res. Logist. Q. **33**(4), 763–774 (1986)
15. Zipkin, P.H.: Foundations of Inventory Management. McGraw-Hill/Irwin, NY (2000)

FIX: The Fear Index—Measuring Market Fear

J. Dhaene, J. Dony, M.B. Forys, D. Linders, and W. Schoutens

Abstract In this paper, we propose a new fear index based on (equity) option surfaces of an index and its components. The quantification of the fear level will be solely based on option price data. The index takes into account market risk via the VIX volatility barometer, liquidity risk via the concept of implied liquidity, and systemic risk and herd behavior via the concept of comonotonicity. It thus allows us to measure an overall level of fear (excluding credit risk) in the market as well as to identify precisely the individual importance of the distinct risk components (market, liquidity, or systemic risk). An additional result, we also derive an upperbound for the VIX.

1 Introduction

The VIX is a key measure of market expectations of near-term volatility conveyed by S&P500 stock index option prices. It is often referred to as the fear index or fear gauge. Since its introduction in 1993, the VIX has been considered by many to be a good barometer of investor sentiment and market volatility. It is a weighted

J. Dhaene (✉) • D. Linders
Department of Accountancy, Finance and Insurance, Katholieke Universiteit Leuven,
Naamsestraat 69, 3000, Leuven, Belgium
e-mail: Jan.Dhaene@econ.kuleuven.be; Daniel.Linders@econ.kuleuven.be

J. Dony
Faculty of Business and Economics, Katholieke Universiteit Leuven, Naamsestraat
69, 3000, Leuven, Belgium

M.B. Forys, and W. Schoutens
Department of Mathematics, Katholieke Universiteit Leuven, Celestijnenlaan
200B, Leuven, Belgium
e-mail: Monika.Forys@wis.kuleuven.be; wim@schoutens.be

M. Cummins et al. (eds.), *Topics in Numerical Methods for Finance*, Springer Proceedings
in Mathematics & Statistics 19, DOI 10.1007/978-1-4614-3433-7_4,
© Springer Science+Business Media New York 2012

blend of prices for a range of options on the S&P500 index. The formula uses a kernel-smoothed estimator that takes as inputs the current market prices for all out-of-the-money calls and puts for the front month and second month expirations. The goal is to estimate the implied volatility of the S&P500 index over the next 30 days. On March 26, 2004, the first-ever trading in futures on the VIX Index was launched on the CBOE Futures Exchange (CFE). Since February 24, 2006, it became possible to trade VIX options contracts.

Actually, the VIX is an indicator of perceived volatility in either direction (including the upside) and, hence, not necessarily bearish for the stocks. Of course it is well documented that volatility and stock returns are negatively correlated.

Next to volatility, there are also other risk or fear factors in the market. Other fear components are, for example, systemic risk, liquidity risk, and counterparty risk. More precisely, in times of heavy distress, besides very high levels of volatility, we typically observe also a drying up of liquidity in the sense that bid and ask spreads widen. When liquidity dries up, one cannot easily unwind positions near theoretical mid prices anymore, but one faces a negative price impact for immediate liquidiations; fire-sale transactions are typically at much lower prices. Furthermore, in such circumstances we also see more herd behavior pointing to a movement of the market into one direction. The later is related to the dependency relationships between traded assets. Finally, the market is well aware of the fact that in stress situations the probability that a counterparty fails is rising. Good indicators of counterpart risk are the credit indices like CDX and iTraxx.

In this paper, we will create a new fear index on the basis of (equity) option surfaces on an index and its components. The quantification of the fear level is, hence, on the basis of option price data only and not on any kind of historical data. The index will take into account market risk, via the VIX volatility barometer, liquidity risk, via the concept of implied liquidity, and finally systemic risk, via the concept of comonotonicity. The index allows us to measure an overall level of fear (excluding credit risk) in the market and to identify exactly the importance of the individual components (market, liquidity, or systemic risk).

As indicated above, the paper will make use of the concept of implied liquidity introduced in [11]. It is based on the fundamental theory of conic finance, in which the one-price theory is abandoned and replaced by a two-price theory giving bid and ask prices for traded assets. The pricing is performed by making use of nonlinear distorted expectations. In essence, the distorted expectation used in [8] is parameterized by one parameter. A high value of this parameter gives rise to a wide bid-ask spread, a low value to a small spread. Given a market bid-ask spread, one can, via reverse engineering (cfr. implied volatility), back out the unique implied parameter to be put into the distortion function to recoup the market spread. This implied parameter is called the implied liquidity parameter. This allows us to measure the degree of liquidity of a certain asset in an isolated manner and to quantify it exactly.

Further, in order to quantify the level of systemic risk in the markets, we make use of the theory of comonotonicity (see [14, 15]). This theory allows us to measure herd behavior, i.e., to which degree the whole market just goes into one direction.

In particular, the comonotonic dependency structure is such that it is driven by one single systemic factor, and so that under a full comonotonic setting, all movements of all the assets are driven by this single factor. By pricing vanilla options on the index, which we see as options on a basket of the underlying components, under the comonotonic dependency structure and comparing these with actual index option prices, we are able to measure how far the observed market prices are away from the fully comonotonic market case. If we are in the theoretical case that the comonotonic gap, i.e., the difference between the comonotonic price and the market price, is zero, we are in a market driven purely by one common factor. If the gap is large, one is closer to a situation where the index components have a fully independent idiosyncratic behavior. The notion *comonotonicity gap* was introduced by Laurence [19]. Comonotonic option prices can be determined via the general procedure presented in [14]. The application of this procedure to the index option case in a market with a finite number of options traded is considered in [7, 19].

This paper is organized as follows. First we elaborate on how exactly to compute the components of the overall market fear index. Then, we bring together the liquidity, the systemic and the volatility component into one overall fear estimate. We do this by taking a weighted sum, where the weights are set that a fear level of 100 represents historically the average case. A number above 100 indicates that the fear is above average; a number below 100 indicates that we are in a fear situation below average. The later is exactly quantified on the basis of a historical study over the period January 2007–October 2009, for which we calculate the fear numbers on a daily basis for the Dow Jones Index. Some key events in the recent credit crisis in that period are clearly identified.

2 Measuring Market Risk via the VIX

In 1993, the Chicago Board Options Exchange (CBOE) introduced a new index, called *VIX*, which aimed at estimating the expected short-term volatility of the S&P100 index over the next 30 days. Initially, VIX used to be an average of eight different implied volatilities calculated from eight at-the-money options of the S&P100. In particular, two ATM calls and two ATM puts were selected for two different maturities (which we will refer to as "near-term" and "next-term" maturities) and the implied volatilities were computed according to a Black–Scholes model [3].

However, model-dependent estimations based on the small range of options inaccurately reflected the real market volatilities. Thus, in September 2003 the *new VIX* has been introduced (see [6]). It is based on a much wider range of options and the underlying index has been replaced by the larger S&P500, which provides stronger correlation with the market than S&P100, as more stocks are involved. Also a model-free approach is used. This model-free approach relies on an volatility estimation developed in [4] combined with an efficient discretization proposed in [18]. There is no model involved and the only requirements are continuity, absence of arbitrage, and Markovian dynamics.

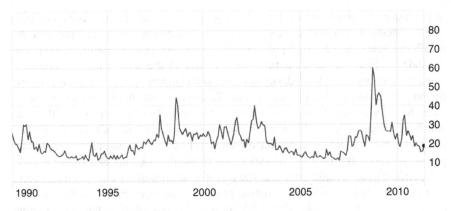

Fig. 1 Historical values of the S&P500 VIX; period 01.1990–12.2010

On March 26, 2004, the first-ever trading in futures on the VIX Index was launched on CFE. Since February 24, 2006, it became possible to trade VIX options contracts. On January 5, 2011, CBOE announced to also VIX-ify individual stocks like APPL, IBM, GS, GOOG,...

The new VIX index is often referred to as the fear index or fear gauge, since its extreme values were achieved during the substantial decreases on the market. As mentioned, the volatility measure aims at estimating the expected short-term volatility of the S&P500 index over the next 30 days. It is calculated using the current market prices of all out-of-the-money options with front month and second month expirations. Values of the VIX index based on S&P500 are depicted in Fig. 1.

Since its introduction, the VIX has been considered by many to be a good barometer of investor's sentiment and market volatility. The VIX typically spikes up when the market falls and goes down when the market goes up. This reflects a natural negative correlation between the VIX and the index returns (see Fig. 2). The VIX thus quantifies the concept of volatility and acts as an effective measure for the expected movements in the next 30 days S&P500 returns.

The VIX index typically fluctuated within a range of 15–30, with an average of 18.97 for the period 04.01.1993–31.12.2007. Due to the worldwide financial crisis in 2008, the VIX reached a value of 80 around November 2008 (see Fig. 1).

The next subsection describes the notion of the model-free estimator for the volatility.

2.1 The Model-Free Estimator for Volatility: The VIX

As mentioned before, the VIX index has not always been calculated in the same way, as in September 2003 the model-free approach based on a wider option surface was introduced. The actual VIX is a weighted blend of prices for a range of options

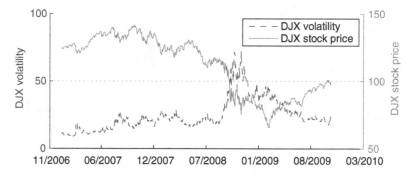

Fig. 2 DJX volatility vs. DJX stock price; period 01.2007–10.2009

on the S&P500 index. The formula uses as inputs the current market prices for all out-of-the-money calls and puts for the front month and second month expirations. The goal is to estimate the volatility of the S&P500 index over the next 30 days.

The following quantity is crucial in the VIX calculation. It gives a model-free estimate for the variance, based only on options with maturity T.

$$\sigma^2 = \frac{2}{T}\sum_i \frac{\Delta K_i}{K_i^2}e^{rT}Q(K_i) - \frac{1}{T}\left(\frac{F}{K_0}-1\right)^2, \tag{1}$$

where

- F is the forward index level. F is determined by first identifying the strike price, K^* at which the absolute difference between the call ($C(K^*,T)$) and put ($P(K^*,T)$) prices is the smallest. Then $F = K^* + e^{rT}(C(K^*,T) - P(K^*,T))$.
- K_0 is the first strike below the forward index level.
- K_i is the strike price of the ith out-of-the money option; a call if $K_i > K_0$ and a put if $K_i < K_0$; both put and call if $K_i = K_0$. The range of the strikes taken into consideration is described in [10].
- ΔK_i is half the difference between the strikes on either side of K_i, i.e.,

$$\Delta K_i = (K_{i+1} - K_{i-1})/2$$

except for the lowest strike, where ΔK is simply the difference between the lowest strike and the next higher strike. Likewise, ΔK for the highest strike is the difference between the highest strike and the next lower strike.

- $Q(K_i)$ is the midpoint of the bid/ask spread for each option with strike K_i. The K_0 put and call prices are averaged to produce a single value.

Here, $C(K,T)$ and $P(K,T)$ denote the respective mid-prices of the call and put options with strike K and maturity T. In contrast, we write $C^{\mathrm{bid}}(K,T)$ and $C^{\mathrm{ask}}(K,T)$ for the bid and ask prices. One can notice that the VIX calculation is very much related to the implementation of a Variance Swap as elaborated on in [5, 13, 22].

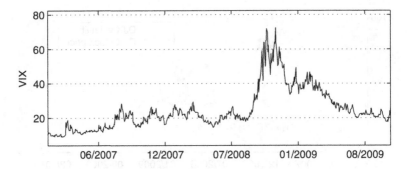

Fig. 3 Dow Jones VIX; period 01.2007–10.2009

For the actual calculation of the VIX index, which is a 30-day forward-looking estimate of the volatility, one needs to compute two variances based on this formula, namely, a first one, σ_1^2, for the near-term options (T_1) and a second one, σ_2^2, for the next-term options (T_2). When the near-term options have less than a week to expiration, the VIX "rolls" to the second and third contract months. The VIX is then an interpolation at the 30 days point, based on values at T_1 and T_2:

$$
\text{VIX}_{30} = 100 \sqrt{ \left\{ T_1 \sigma_1^2 \underbrace{\left[\frac{N_{T_2} - N_{30}}{N_{T_2} - N_{T_1}} \right]}_{x_1} + T_2 \sigma_2^2 \underbrace{\left[\frac{N_{30} - N_{T_1}}{N_{T_2} - N_{T_1}} \right]}_{1-x_1} \right\} \frac{N_{365}}{N_{30}} }, \tag{2}
$$

where:

- N_{T_1} = number of minutes to settlement of the near-term options (i.e., with maturity T_1).
- N_{T_2} = number of minutes to settlement of the next-term options (i.e., with maturity T_2).
- N_{30} = number of minutes in 30 days.
- N_{365} = number of minutes in a 365-day year.

The VIX methodology was historically introduced on the S&P500 options and later applied to several other indices, stocks, and assets. The VIX calculations conducted in this research will in contrast be based on the Dow Jones Industrial Average Index (DJX). Figure 3 shows the DJX VIX estimate in the period between January 2007 and October 2009.

One can observe a mean-reverting behavior of the VIX. In the period preceding the credit crisis, the VIX underwent a rapid growth and went from a value of 20 up to a value of more than 70 in a timespan of a few weeks only. The S&P500 VIX actually went up even to 80. Hence, at that point, at the heat of the financial crisis, the market was expecting unusually large movements of the stocks. We also remark that in 2010, the DJX VIX has come down from its highest levels back under the 30 level again. The average Dow Jones VIX level for this period is calculated as 24.66.

3 Measuring Liquidity Risk via the Implied Liquidity

In the previous section, it was shown that volatility levels can give us an indication of the nervousness of the market conditions. Liquidity is another important measure, which reflects an asset's ability to be sold. High bid-ask spreads characterize illiquid products, whereas liquidity implicates a smaller spread. However, it is very difficult to measure liquidity in an isolate manner. Bid-ask spreads can move around in a nonlinear manner if spot or volatility moves, without a change in liquidity.

In the sequel, we will discuss the concept of implied liquidity, which in a unique and fundamentally founded way isolates and quantifies the liquidity risk in financial markets. This concept was already proposed in [11] and is based on the theory of conic finance, in which the one-price theory is abandoned and the two-price market is employed.

3.1 Conic Finance—Bid and Ask Pricing

In this section, we summarize the basic conic finance techniques needed to calculate the implied liquidity parameter related to a vanilla option position. For more background, see [8, 9, 21]. Conic finance uses distortion functions to calculate distorted expectations. In [11], a distortion function from the minmaxvar family parameterized by a single parameter $\lambda \geq 0$ as in (3) is chosen.

$$\Phi(u;\lambda) = 1 - \left(1 - u^{\frac{1}{1+\lambda}}\right)^{1+\lambda}. \tag{3}$$

Hereafter, we will employ distorted expectations to calculate bid and ask prices. The prices arise from the theory of acceptability. A risk X is said to be acceptable (notation: $X \in \mathscr{A}$) if

$$E_Q[X] \geq 0 \text{ for all measures } Q \text{ in a convex set } \mathscr{M}.$$

The convex set $\mathscr{M}$ contains the supporting measures, which can be seen as a kind of test measures under which the cash flow X needs to have positive expectation to deliver acceptability. Under a larger set $\mathscr{M}$, one has a smaller set of acceptable risks, because there are more underlying tests to be passed.

Operational cones were defined by Cherney and Madan [8] and depend solely on the distribution function $G(x)$ of X and a distortion function Φ. Here $X \in \mathscr{A}$ if the distorted expectation is non-negative. More precisely, the distorted expectation with respect to the distortion function Φ (we use the one given in (3) but other distortion functions are also possible), of a random variable X with distribution function $G(x)$, is defined as

$$\mathrm{de}(X;\lambda) = \int\limits_{-\infty}^{+\infty} x \mathrm{d}\Phi(G(x);\lambda). \tag{4}$$

Note that if $\lambda = 0$, $\Phi(u;0) = u$, and, hence, $de(X;0) = E[X]$ is equal to the original expectation.

The ask price of payoff X is determined by

$$\text{ask}(X) = -\exp(-rT)de(-X;\lambda).$$

This formula is derived by noting that the cash flow of selling X at its ask price is acceptable in the relevant market, that is $\text{ask}(X) - X \in \mathscr{A}$. Similarly, the bid price of payoff X is determined as

$$\text{bid}(X) = \exp(-rT)de(X;\lambda).$$

Here the cash flow of buying X at its bid price is acceptable in the relevant market: $X - \text{bid}(X) \in \mathscr{A}$.

One can prove that the bid and ask prices of a positive contingent claim X with distribution function $G(x)$ can be calculated as

$$\text{bid}(X) = \exp(-rT) \int_0^{+\infty} x \, d\Phi(G(x);\lambda), \tag{5}$$

$$\text{ask}(X) = \exp(-rT) \int_{-\infty}^0 (-x) \, d\Phi(1 - G(-x);\lambda). \tag{6}$$

Suppose now that we are given a market bid and ask price for a European call. We can then calculate the mid price of that call option, as the average of the bid and ask prices. Out of this mid price we calculate the implied Black–Scholes volatility, to calculate the conic bid and ask prices (using the implied volatility as parameter). Under the Black–Scholes framework, this comes down to the following calculations for an European call option with strike K and maturity T. The distribution of the call payoff random variable to be used in (5) and (6) is in this case given by

$$G(x) = 1 - \text{N}\left(\frac{\log(S_0/(K+x)) + (r - q - \sigma^2/2)T}{\sigma\sqrt{T}}\right), \qquad x \geq 0,$$

where S_0 is the current stock price, r the risk-free rate, and q the dividend yield. Further, N is the cumulative distribution function of the standard normal law and σ is the implied volatility determined on the basis of the mid price. For $x < 0$, $G(x) = 0$, since the payoff is a positive random variable. The above closed-form solution for $G(x)$ in combination with (5) and (6) gives rise to very fast and accurate calculations of the bid and ask prices.

The parameter λ in (5) and (6), which fits the bid-ask spread around the mid price the best, is then called *the implied liquidity parameter*. The smaller the implied liquidity parameter, the more liquid the underlying and the smaller the bid-ask spread. In the extremal case where the implied liquidity parameter equals 0, the bid price coincides with the ask price, and we are back in the one-price framework.

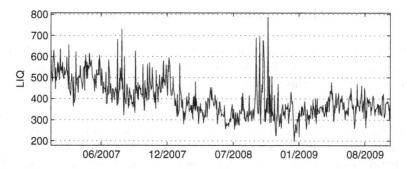

Fig. 4 Dow Jones LIQ; period 01.2007–10.2009

3.2 Measuring Liquidity with LIQ

It is well-known that a distressed market suffers from drying up liquidity. In order to measure the liquidity risk, we propose a measure based on implied liquidity, which we will call LIQ.

We denote by LIQ_j the 30-days implied liquidity of company j, calculated from the near- and next-term implied liquidities: $\lambda_j^*(T_1)$ and $\lambda_j^*(T_2)$. We compute it using the same weights as in the VIX methodology. $\lambda_j^*(T_i), i = 1, 2$ itself is calculated as an average of all the individual implied liquidities of all nonzero bid call and put options of company j. Hence, LIQ_j of the j-th company is given by:

$$LIQ_j = x_1 \lambda_j^*(T_1) + (1 - x_1)\lambda_j^*(T_2).$$

In the same way we calculate the implied liquidity LIQ_{DJX} of the index. This combination of near- and next-term liquidities provides a *short-term forward-looking implied liquidity*.

The overall liquidity index for a particular day is defined as:

$$LIQ = \frac{1}{2}LIQ_{DJX} + \frac{1}{2n} \sum_{j=1}^{n} LIQ_j.$$

In Fig. 4, the market liquidity estimation based on the DJX index and all the 30 underlying stocks is presented. We clearly observe that LIQ is not constant over time and apparently exhibits a mean-reverting behavior. Recent work investigates this stochastic liquidity behavior more in depth, see [2].

The long-run average of the implied liquidity of the data set in the period between January 2007 and October 2009 equals 412 bp. The highest value of the LIQ parameter, 1260 bp, was reached on the 24th of October 2008. Around this day, several European banks were rescued by government intervention.

4 Measuring Herd-Behavior via the Comonotonicity

In this section, we introduce a third ingredient contributing to the general panic level in the market, namely, herd behavior. This notion refers to the tendency of one decision maker to take his decisions in accordance with those of a whole group of decision makers, whether or not these decisions are rational.

When the market is more agitated, it is therefore not unusual to observe a stronger herd behavior pointing to a systemic movement of the market into one direction. Measuring the risk of herd behavior is not necessarily straightforward, as it relies on human's reactions in specific circumstances, which are not easy to quantify. However, the herd phenomenon in the financial market is also essentially related to the dependency relationship between the traded assets.

Inspired by [16], we propose a measure for herd behavior in the market based on the concept of comonotonicity. While in [16], the implied variance of the index price is compared with its comonotonic version and its ratio is called the HIX, we propose to compare the VIX index by its comonotonic version. From a methodological point of view, this is very similar to the HIX approach. Therefore, for a more profound study of the underlying methodology, we refer to [16]. For an overview of the theory of comonotonicity and its applications in insurance and finance, we refer to [14,15]. For further work on the applications of this theory in an option pricing framework, we refer to [1,7,20,23]. A recent overview of the literature on financial applications of the theory of comonotonicity is given in [12].

4.1 Comonotonicity

In this section, we will summarize basic concepts of comonotonicity theory in relationship with the dependency structure between the underlyings in a basket of assets. We start with stating some main results concerning comonotonicity theory. Definitions, results, and detailed proofs can be found in [7, 14]. Let us consider n different stocks $i = 1,\dots,n$ with corresponding stock price processes $\{S_i(t), t \geq 0\}$. These stocks form an index (or basket) consisting of a combination of a certain amount w_i of stock i, where $w_1, w_2, \dots, w_n$ is a series of upfront fixed positive weights. We denote by $\{\mathscr{S}(t), t \geq 0\}$ the price process of the index calculated as the weighted sum of the n underlying stock price components, i.e.,

$$\mathscr{S}(t) = \sum_{i=1}^{n} w_i S_i(t), \quad t \geq 0.$$

In our example, we will use $n = 30$, and work with the 30 components $(S_i, i = 1,\dots,30)$ of the DJX index $(\mathscr{S})$.

Suppose there exists an option market of vanilla calls and puts on the individual stocks $i = 1,\dots,n$, as well as on the index. We shall denote by $C_i(K,T)$ and $P_i(K,T)$

the prices of a European call option and European put option respectively on stock i with strike K and maturity T. In the same way, we write $\mathscr{C}(K,T)$ and $\mathscr{P}(K,T)$ for the option prices on the index.

Recall that the payoff of a European call with strike K and maturity T on the index is given by $(\mathscr{S}(T) - K)^+ = (\sum_{i=1}^n w_i S_i(T) - K)^+$. In order to compute the price of this call option, one would actually need to have full knowledge about the dependency structure of the underlying stocks. This information is usually not known; however, one can always find an optimal upper bound of $\mathscr{C}(K,T)$ by taking a linear combination of observable call option prices $C_i(K,T)$, and which corresponds to the case when the stock price vector is comonotonic. This leads us to the definition of comonotonic vectors.

Definition 1 (Comonotonic vector). Let $Y_1,\ldots,Y_n$ be arbitrary random variables and let U be a uniformly distributed random variable on the unit interval. We say that the random vector $\mathbf{Y} = (Y_1,\ldots,Y_n)$ is comonotonic if

$$\mathbf{Y} \stackrel{d}{=} \left(F_{Y_1}^{[-1]}(U),\ldots,F_{Y_n}^{[-1]}(U) \right),$$

where $\stackrel{d}{=}$ stands for equal in distribution, and $F_Y^{[-1]}(u) = \inf\{x \in \mathbb{R} : F_Y(x) \geq u\}$ (and $\inf \emptyset = +\infty$ by convention).

The comonotonic vector is driven by only one single factor (U)—the systemic risk. Now, for any random vector $\mathbf{X} = (X_1,\ldots,X_n)$, we can define the so-called *comonotonic counterpart* of $\mathbf{X}$. It is denoted by $\mathbf{X}^c$ and is defined as

$$\mathbf{X}^c \equiv (X_1^c,\ldots,X_n^c) \stackrel{d}{=} \left(F_{X_1}^{[-1]}(U),\ldots,F_{X_n}^{[-1]}(U) \right).$$

In this context, we define the comonotonic index price process as:

$$\mathscr{S}^c(t) = \sum_{i=1}^n w_i S_i^c(t), \quad t \leq T,$$

where $S^c(t) = (S_1^c(t),\ldots,S_n^c(t))$ is the comonotonic counterpart of the stock price vector $S(t) = (S_1(t),\ldots,S_n(t))$. In analogy to the regular index call price, we will denote

$$\mathscr{C}^c(K,T) = e^{-rT} \mathbb{E}_Q \left[(\mathscr{S}^c(T) - K)^+ \right] \tag{7}$$

for the comonotonic call value. Note that the comonotonic version incorporates perfect herd behavior, and index call options under perfect herd behavior should intuitively be more expensive, since each index component moves in the same direction and, hence, the index exhibits a higher volatility. From now on, to avoid unnecessary overloading of notation, we will omit writing "(t)" whenever there is no confusion possible. In particular, we will write $\mathscr{S}^c \equiv \mathscr{S}^c(T)$ and $S_i(T) \equiv S_i$.

4.2 Comonotonic Upper Bound

In this section, it will be shown how to derive an upper bound for a call option on the index in terms of call options on the individual stocks. For details, we refer to [7].

In fact comonotonicity theory implies, that it is always possible to bound the index option price $\mathscr{C}(K,T)$ from above, namely, with the price of the comonotonic index call price $\mathscr{C}^c(K,T)$. To do so, we first have to specify the comonotonic distribution $F_{\mathscr{S}^c}$.

Theorem 1 (Comonotonic distribution). *The distribution function of the comonotonic index price process is given for any $x \in (F_{\mathscr{S}^c}^{-1+}(0), F_{\mathscr{S}^c}^{-1}(1))$ as*

$$F_{\mathscr{S}^c}(x) = \sup\left\{ p \in [0,1] : \sum_{i=1}^{n} w_i F_{S_i}^{[-1]}(p) \leq x \right\}, \tag{8}$$

where for each $0 < \alpha \leq 1$ the alpha-inverse of $F_{\mathscr{S}^c}$ is given by

$$F_{\mathscr{S}^c}^{[-1(\alpha)]}(p) = \sum_{i=1}^{n} w_i F_{S_i}^{[-1(\alpha)]}(p), \quad 0 < \alpha \leq 1,$$

and the alpha-inverse is defined as $F_Y^{[-1(\alpha)]}(u) = \alpha F_Y^{[-1]}(u) + (1-\alpha)F_Y^{[-1+]}(u), 0 < \alpha < 1$, with $F_Y^{[-1+]}(u) = \sup\{x \in \mathbb{R} : F_Y(x) \leq u\}$ (with $\sup \emptyset = -\infty$, by convention).

We are now able to calculate the expected payoff of a call option on the comonotonic basket, as it is shown in the following theorem; for a proof see [7].

Theorem 2 (Comonotonic index option price). *Let $\mathscr{S}^c$ be the comonotonic price process of an index as above. Then*

$$\mathbb{E}\left[(\mathscr{S}^c - K)^+\right] = \sum_{i=1}^{n} w_i \mathbb{E}\left(S_i - F_{S_i}^{[-1(\alpha)]}(F_{\mathscr{S}^c}(K))\right)^+, \tag{9}$$

where $\alpha \in [0,1]$ must be chosen in such a way that $F_{\mathscr{S}^c}^{[-1(\alpha)]}(F_{\mathscr{S}^c}(K)) = K$, or equivalently (by the additivity property of comonotonic quantiles),

$$\sum_{i=1}^{n} w_i F_{S_i}^{[-1(\alpha)]}(F_{\mathscr{S}^c}(K)) = K.$$

Consequently, the comonotonic index option price is given by

$$\mathscr{C}^c(K,T) = \sum_{i=1}^{n} w_i C_i\left(F_{S_i}^{[-1(\alpha)]}(F_{\mathscr{S}^c}(K)), T\right).$$

Essentially, the above expression tells us that the price of a call option with strike K and maturity T on the index under the comonotonic setting equals a weighted sum of call prices on the index components. The weights are identical to the ones used for the index composition and the maturities are identical as well. The strikes are given by

$$K_i^* = F_{S_i}^{[-1(\alpha)]} (F_{\mathscr{S}^c}(K)). \tag{10}$$

To determine these strikes, we need to know the distribution functions of S_i, $F_{S_i}(x)$ and the distribution function of the comonotonic index $F_{\mathscr{S}^c}$. The distribution function of $F_{S_i}(x)$ can be extracted from the option surface of stock i:

$$F_{S_i}(x) = 1 + e^{rT} \frac{\partial C(K,T)}{\partial K} |_{K=x+}, \quad x > 0. \tag{11}$$

Given the marginal distribution functions, the comonotonic distribution function can be calculated using (8). Note that both (8) and (11) can be calculated in a model free way using only option price data.

The above formula (11) is, however, only valid if call options are available for any strike. In the real world this is not the case, and only a finite number of call prices are available for a given maturity. Therefore, in [17] (see also [7]) one proposes to approximate $F_{S_i}(x)$ by a piecewise constant function $\bar{F}_{S_i}(x)$ defined as

$$\bar{F}_{S_i}(K_{i,j}) = 1 + e^{rT} \frac{C_i(K_{i,j+1},T) - C_i(K_{i,j},T)}{K_{i,j+1} - K_{i,j}}, \tag{12}$$

where $K_{i,j}$, $j = 1, \ldots, m_i$, are the traded strikes for the underlying stock i. Finally, we have that

$$\bar{F}_{\mathscr{S}^c}(x) = \sup \left\{ p \in [0,1] : \sum_{i=1}^n w_i \bar{F}_{S_i}^{[-1]}(p) \leq x \right\}.$$

Having all the formulas at hand, we can define the comonotonic upper bound: for all strikes K in the support of $F_{\mathscr{S}^c}$ we can bound the index option price by

$$\mathscr{C}(K,T) \leq \sum_{i \in \mathscr{N}_K} w_i C_i(K_{i,j_i},T) + \sum_{i \notin \mathscr{N}_K} w_i \left\{ \alpha_K C_i(K_{i,j_i},T) + (1 - \alpha_K)C_i(K_{i,j_i+1},T) \right\}, \tag{13}$$

where $j_1, \ldots, j_n$ and $\mathscr{N}_K$ are (sets of) indices depending on $F_{\bar{\mathscr{S}}^c}(K)$, and where α_K is a function of *observed* call option prices $C_i(K_{i,j},T)$ only. In particular, we have that

1. The indices j_i are such that $\bar{F}_{S_i}(K_{i,j_i-1}) < F_{\bar{\mathscr{S}}^c}(K) \leq \bar{F}_{S_i}(K_{i,j_i})$
2. $\mathscr{N}_K = \{i \leq n : F_{\mathscr{S}^c}(K) \neq \bar{F}_{S_i}(K_{i,j_i})\}$
3. α_K is any number satisfying $\bar{F}_{\mathscr{S}^c}^{[-1(\alpha_K)]}(F_{\bar{\mathscr{S}}^c}(K)) = K$, or equivalently,

$$\sum_{i=1}^{n} w_i \bar{F}_{S_i}^{[-1(\alpha_K)]}(F_{\mathscr{G}^c}(K)) = K. \tag{14}$$

The comonotonic upper bounds can be computed for and on the basis of put options as well [20]. There is only one formula that requires adaptation, i.e., the empirical distribution function $\bar{F}_{S_i}(K_{i,j})$. The expression in (12) then becomes

$$\bar{F}_{S_i}(K_{i,j}) = e^{rT} \frac{P_i(K_{i,j+1}, T) - P_i(K_{i,j}, T)}{K_{i,j+1} - K_{i,j}}. \tag{15}$$

The comonotonic upper bound for the index put option is then the analogue of (13), and can be formulated in the following way: For all strikes K in the support of $F_{\mathscr{G}^c}$ we can bound the index put option price by

$$\mathscr{P}(K,T) \leq \sum_{i \in \mathscr{N}_K} w_i P_i(K_{i,j_i}, T) + \sum_{i \notin \mathscr{N}_K} w_i \{\alpha_K P_i(K_{i,j_i}, T) + (1 - \alpha_K) P_i(K_{i,j_i+1}, T)\},$$

where $j_1, \ldots, j_n$ and $\mathscr{N}_K$ are (sets of) indices depending on $F_{\mathscr{G}^c}(K)$, and where α_K is a function of *observed* put option prices $P_i(K_{i,j}, T)$ only.

4.3 The Comonotonicity Ratio

Knowing both the index option price $\mathscr{C}(K,T)$ for a certain strike K and maturity T and its upper bound $\sum_{i=1}^{n} w_i C_i(K_i^*, T) = \mathscr{C}^c(K,T)$, one can compare both values to measure how far one is away from the fully comonotonic situation. On the basis of this, [19] proposes the so-called *comonotonicity gap*, which compares the market's price with the perfectly comonotonic price by means of their ratio. We work with the related *comonotonicity ratio:*

$$\rho_{\text{call}}(K,T) = \frac{\mathscr{C}(K,T)}{\mathscr{C}^c(K,T)}. \tag{16}$$

Alternatively, in order to have a more robust and overall comonotonicity measure based on similar ideas as the one proposed in [16], we VIX-ify the above comonotonicity ratio by replacing call and put option quotes in the VIX formula by their comonotonic upper bound. More precisely, in (1) we replace $Q(K_i)$ by its comonotonic upper bound $Q^c(K_i)$, calculated according to the formulas above. This results in the following formula:

$$\sigma_{\text{com}}^2(T) = \frac{2}{T} \sum_i \frac{\Delta K_i}{K_i^2} e^{rT} Q^c(K_i) - \frac{1}{T} \left(\frac{F}{K_0} - 1\right)^2.$$

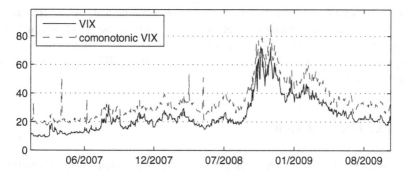

Fig. 5 Dow Jones *VIX* and VIX^c; period 01.2007–10.2009

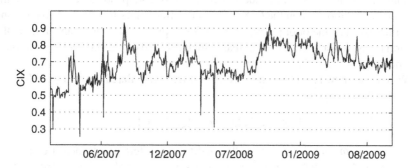

Fig. 6 Dow Jones *CIX*; period 01.2007–10.2009

In this way and using the same interpolation on the 30-days point (using the front and next month maturities), we derived a *comonotonic VIX* (VIX^c), which is a market-implied upper bound for the VIX (Fig. 5):

$$\text{VIX}^c = 100 \sqrt{\left\{ T_1 \sigma_{\text{com}}^2(T_1) \underbrace{\left[\frac{N_{T_2} - N_{30}}{N_{T_2} - N_{T_1}}\right]}_{x_1} + T_2 \sigma_{\text{com}}^2(T_2) \underbrace{\left[\frac{N_{30} - N_{T_1}}{N_{T_2} - N_{T_1}}\right]}_{1-x_1} \right\} \frac{N_{365}}{N_{30}}}.$$

We graph in Fig. 5, the DJX VIX and its upperbound VIX^c.

Finally, we define the *comonotonicity VIX ratio*, baptized CIX, as the ratio of the regular VIX and the comonotonic VIX, i.e.,

$$\text{CIX} = \rho_{\text{VIX}} = \frac{\text{VIX}}{\text{VIX}^c}. \qquad (17)$$

The CIX can be used as a measure for systemic risk and herd behavior (Fig. 6). The closer its value is to the closer we are to the comonotonic situation and the more systemic risk or herd behavior there is in the market. Perfect herd behavior is reached when CIX = 1. Hence, the ratio gives us a simple and convenient way to measure how much herd behavior is present in the market, and thus to quantify the systemic risk on the basis of traded option information.

Again, the credit crisis is clearly visible around October 2008 as well as financial issues during the summer of 2007.

5   FIX: The Fear Index

In the previous chapters, we have proposed measures for quantifying several types of risks in the market. As such, we have discussed the VIX as a (model-free) estimate for market risk, introduced LIQ as (model-dependent) estimate for the liquidity and elaborated on the CIX as a (model-free) measure of herd behavior and systemic risk. The major objective of these developments was the establishment of a general measure for overall market fear, which is based on the three aforementioned components, combined in a particular way.

We call our new fear index the FIX. And, FIX is calculated out of VIX, LIQ, and CIX as follows:

$$FIX = \omega_1 VIX + \omega_2 LIQ + \omega_3 CIX,$$

where $\omega_1, \omega_2, \omega_3$ are the weights allocated to the different risk measures in such a way that the contribution of each risk is proportional to its contribution to the "average fear situation." Based on our previous results and calculations, the respective average values for the DJX Index over the period 2007–2009 are estimated as follows:

$$\widetilde{VIX} = 24.66\%, \quad \widetilde{LIQ} = 400.65\,bp, \quad \widetilde{CIX} = 69.16\%.$$

We now define the weights ω_1, ω_2, and ω_3 in such a way that

$$0.25\omega_1 = 0.04\omega_2 = 0.7\omega_3 = \frac{100}{3},$$

where the choices 25 %, 400 bp and 70 % are settled in accordance with the obtained averages. Applying these values leads to the contribution of each component in the FIX:

$$\omega_1 = 133.33, \quad \omega_2 = 833.33, \quad \omega_3 = 47.62.$$

These choices will then lead to a fear measure FIX having an average level of 100. A value FIX > 100 will reflect a market with a fear level above average, whereas a value FIX < 100 expresses less fear in the market than average. Application of these values in the calculation of the FIX gives the following plot of the Fear Index FIX as shown in Fig. 7.

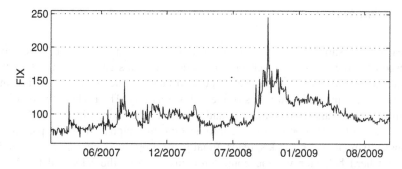

Fig. 7 Dow Jones Fear Index; period 01.2007–10.2009

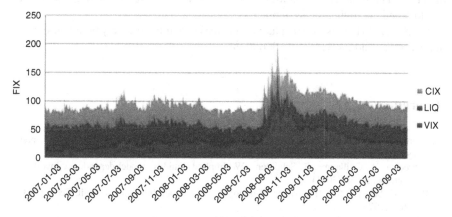

Fig. 8 Dow Jones Fear Index components; period 01.2007–10.2009

The different market fear components are shown in Fig. 8. The pattern of Fig. 8 clearly reflects the financial problems of the past few years. For instance, the peak on June, 7, 2007 coincides with the announcement by Bear Stearns to the investors that it is suspending redemptions from its HighGrade Structured Credit Strategies Enhanced Leverage Fund. Two months later, in August 2007, the FIX peaks again. In this case, it goes along with bankruptcy of American Home Mortgage Investment Corp. (NYSE: AHM) on August 6, 2007. The following days numerous quantitative long/short equity hedge funds suddenly began experiencing unprecedented losses. As such, on August 9, 2007, BNP Paribas SA, France's largest bank, suspended three investment funds because it could not "fairly" value their holdings after the U.S. subprime mortgage losses roiled credit markets. From 10 August 2007 on, the Central Banks around the world started injecting funds into markets as a response to an undesired and unwelcome spike in short-term rates. As a last example, we mention the huge increase in the Fear Index in October 2008 revealing the global financial crisis.

5.1 Conclusions

"Market fear" should be measured by several factors. In this research we have focused on three of them, which in our opinion, have a significant impact on the overall market fear level. First, we propose to take into account market risk and nervousness, expressed it in terms of the index volatility. The higher the volatility, the more market uncertainty there is and the wider the swings in the market can occur. Secondly, we propose to take into account the implied liquidity parameter intrinsically related to bid-ask spreads. Finally, we propose to measure the systemic risk and herd behavior via the comonotonicity ratio of the VIX and the VIX-ified comonotonic upperbound. In a systemic crisis, all assets move into the same direction. The more comonotonic-like behavior we observe the more assets move together and the higher the systemic risk.

We presented the historical values of the market fear index solely based on vanilla index options and individual stock options.

References

1. Albrecher, H., Dhaene, J., Goovaerts, M., Schoutens, W.: Static hedging of Asian options under Lévy models: the comonotonicity approach. J. Derivatives **12**(3), 63–72 (2005)
2. Albrecher, H., Guillaume, F., Schoutens, W.: Implied liquidity: model dependency investigation. Internal report (2011)
3. Black, F., Scholes, M.: The pricing of options and corporate liabilities. J. Polit. Econ. **81**(3), 637–654 (1973)
4. Britten-Jones, M., Neuberger, A.: Option prices, implied price processes and stochastic volatility. J Finance **55**, 839–866 (2000)
5. Carr, P., Madan, D.: Towards a theory of volatility trading. In: Jarrow, R. (ed.) Volatility, risk publications, pp. 417—427 (1998)
6. Carr, P., Wu, L.: A tale of two indices. J. Derivatives **13**(3), 13–29 (2006)
7. Chen, X., Deelstra, G., Dhaene, J., Vanmaele, M.: Static super-replicating strategies for a class of exotic options. Insur. Math. Econ. **42**(3), 1067–1085 (2008)
8. Cherny, A., Madan, D.B.: New measures of performance evaluation. Rev. Financ. Stud. **22**, 2571–2606 (2009)
9. Cherny, A., Madan, D.B.: Markets as a counterparty: an introduction to conic finance. Int. J. Theor. Appl. Finance **13**(8), 1149–1177 (2010)
10. Chicago Board Options Exchange, Inc.: The CBOE volatility index – VIX. White paper (2009)
11. Corcuera, J.M., Guillaume, F., Madan, D.B., Schoutens, W.: Implied liquidity—towards liquidity modeling and liquidity trading. International Journal of Portfolio Analysis and Management (2012), to appear
12. Deelstra, G., Dhaene, J., Vanmaele, M.: An overview of comonotonicity and its applications in finance and insurance. In: Oksendal, B., Nunno, G. (eds.) Advanced Mathematical Methods for Finance, Springer, Germany (Heidelberg) (2010)
13. Demeterfi, K., Derman, E., Kamal, M., Zhou, J.: More than you ever wanted to know about volatility swaps, Goldman Sachs quantitative strategies research notes (1999)
14. Dhaene, J., Denuit, M., Goovaerts, M.J., Kaas, R., Vyncke D.: The concept of comonotonicity in actuarial science and finance: theory. Insur. Math. Econ. Issue 1 (**31**):3–33 (2002)

15. Dhaene, J., Denuit, M., Goovaerts, M.J., Kaas, R., Vyncke, D.: The concept of comonotonicity in Actuarial science and finance: applications, Insur. Math. Econ. 133–161 (2002)
16. Dhaene, J., Linders, D., Schoutens, W., Vyncke, V.: Insurance: Mathematics and economics, **50**(3), 357–370 (2012)
17. Hobson, D., Laurence, P., Wang, T.H.: Static-arbitrage upper bounds for the prices of basket options. Quant. Finance **5**(4), 329–342 (2005)
18. Jiang, G.J., Tian, Y.S.: Model-free implied volatility and its information content. Rev. Financ. Stud. **18**(4), 1305–1342 (2005)
19. Laurence, P.: Hedging and pricing of generalized spread options and the market implied comonotonicity gap, Presented at the Workshop and Mid-Term Conference on Advanced Mathematical Methods for Finance, Vienna University (2007)
20. Linders, D., Dhaene, J., Schoutens, W.: Some results on comonotonicity based upper bounds for index options. Working paper (2011)
21. Madan, D.B., Schoutens, W.: Conic financial markets and corporate finance. Int J Theor Appl Finance, **14**(5), 587–610
22. Neuberger, A.: Volatility trading, London Business School working paper (1990)
23. Simon, S., Goovaerts, M., Dhaene, J.: An easy computable upper bound for the price of an arithmetic Asian option. Insur. Math. Econ. **26**(2), 175–183 (2001)

American Option Pricing Using Simulation and Regression: Numerical Convergence Results

Lars Stentoft

Abstract Recently, simulation methods combined with regression techniques have gained importance when it comes to American option pricing. In this paper, we consider such methods and we examine numerically their convergence properties. We first consider the Least Squares Monte-Carlo (LSM) method of (Longstaff and Schwartz, Rev Financ Stud, 14:113–147, 2001) and report convergence rates for the cross-sectional regressions as well as for the estimated price. The results show that the method converges fast, and this holds even with multiple early exercises and with multiple stochastic factors as long as the payoff function is smooth. We also compare the convergence rates to those obtained when using the related methods proposed by (Carriere, Insur Math Econ, 19:19–30, 1996; Tsitsiklis JN, Van Roy, IEEE Trans Neural Network, 12(4):694–703, 2001). The results show that the price estimates from the latter methods converge significantly slower in the multi-period situation.

1 Introduction

Although it was believed for a long time that Monte Carlo simulation could be used to price European-style derivatives only, recently it has become clear that this is not the case and it is now well known that simulation methods can be used to price options with early exercise features as well. The challenge is that in order to price such options an optimal exercise strategy has to be determined, and this strategy depends on a series of conditional expectations which are difficult to calculate using simulation methods. In the work by [7, 17, 26] methods for approximating the

L. Stentoft (✉)
Department of Finance, HEC Montreal, 3000 chemin de la Cote-Sainte-Catherine,
Montreal (Quebec) Canada H3T 2A7
e-mail: lars.stentoft@hec.ca

M. Cummins et al. (eds.), *Topics in Numerical Methods for Finance*, Springer Proceedings in Mathematics & Statistics 19, DOI 10.1007/978-1-4614-3433-7_5,
© Springer Science+Business Media New York 2012

conditional expectations are provided that use cross-sectional regressions together with Monte Carlo simulation. Other related work includes the bundling or state space partitioning methods of [2,25], both of which can be considered as very crude regression methods essentially using dummy variables for the partitions of the state space [24].[1]

The methods of [7,17,26] are often considered to be very similar; however, there are important and fundamental differences as shown in [23]. In particular, the Least Squares Monte Carlo approach (henceforth the LSM method) of [17] leads to less biased results particularly when considering multiple early exercises. Moreover, it is probably fair to say that among these methods the LSM method has received the most attention. For example, a search using scholar.google.com early June, 2011, revealed 218 citations of [7], 250 citations of [26], and an astonishing 1,295 citations of [17]. The LSM method has been applied in various settings and the method has proven to be very flexible. For example, it has been used to price life insurance contracts and real-estate derivatives as well as for valuation of real options such as gas storage, mine expansion decisions, and timber harvest contracts. A general finding in many of these applications is that in most settings the LSM method works surprisingly well.

Theoretically, the mathematical foundation for the use of the LSM method is provided in, e.g., [22]. Additional results can be found in the original paper by [17] and in [9], among others. However, it is important to keep in mind that these results are limiting results only, and in any actual application of the method several choices will have to be made, which may influence the performance of the method and hence affect the results obtained. To be specific, a choice has to be made in terms of the number of paths to be used in the simulation and in terms of the number of regressors as well as which type of regressors to be used in the cross-sectional regressions. Moreover, because computational time is increasing in both these dimensions a trade-off often exists. The results in [22] do, however, provide some guidance about these choices as results are derived for the convergence rate of the estimated conditional expectations in the LSM algorithm.

In this paper, we examine the issue of convergence of this type of algorithms along several dimensions. First of all, we provide results which confirm the theoretical rates derived in the simple two-period case with one underlying stochastic factor. Secondly, we present results illustrating that in this situation the convergence rate can be increased towards the best possible rate by optimally picking the rate at which the number of paths is increased as a function of the number of regressors. Thirdly, we consider the situation with several underlying stochastic factors. The results show that for options with smooth payoff functions, e.g., geometric or arithmetic average options, convergence rates of the same order are found in the two-dimensional case. However, for options with payoffs given by the maximum

---

[1]Though the methods proposed in [5,6] also use simulation, they do not directly rely on regression techniques for the approximation. Instead, the first method uses additional subsampling whereas the latter requires information about the transition densities to approximate the continuation value.

or the minimum of two assets convergence rates are estimated to be lower. The reason for this is that these options have payoffs which are non-smooth. Finally, we consider the situation with multiple early exercises. The results show that even with 25 exercise times convergence of the estimated conditional expectation is obtained. Moreover, the estimated order of convergence remains very close to what is obtained with only two steps.

Next, though the main object of interest in the theoretical literature is convergence of the estimated conditional expectations, it is of equal importance to analyze the convergence rate of the estimated option price. However, these rates are generally not available theoretically. This paper contains numerical results for the convergence rates of the American price estimates for each of the cases considered above. We find, perhaps somewhat surprising, that the estimated price often converges with a rate close to the best possible rate. To be specific, the numerical results show that the estimated order of convergence is insignificantly different from the optimal one could hope for given by the order of convergence of the corresponding simulated European price estimate. This holds for the benchmark situation as well as when the number of early exercises is increased. It also holds in the two-dimensional case in general, though the option on the minimum of two assets is an exception. In particular, though the estimated order of convergence is still large in numerical value for this option, it is significantly lower than the optimal one.

Finally, we consider two situations which are not covered by the existing theory. The first is the case when the polynomial family used in the cross-sectional regressions is changed. We examine four alternative polynomial families: Laguerre, Hermite, and Chebyshev polynomials of the first and second kind. The results show that care has to be taken in terms of the choice of regressors. In particular, we show numerically that the LSM method does not appear to converge when Laguerre or Hermite polynomials are used. Secondly, we consider the case when all the paths are used in the cross-sectional regressions instead of only the in-the-money paths as proposed by [17]. The use of all the paths is, e.g., necessary in applications of the methods proposed by [7, 26]. The results show that, though convergence of the estimated conditional expectation remains fast, the rate at which the price converges deteriorates as the number of early exercises is increased when the methods of [7,26] are used. For example, when 25 exercise times are considered the convergence rate of the price estimate obtained with these methods is estimated to be significantly lower than the rate obtained when using the LSM method.

The rest of the paper is outlined as follows: In Sect. 2, we discuss how American options can be priced using simulation and regression techniques. In Sect. 3, the convergence rate for a set of artificial American put options is examined and we document the possibility of speeding up this rate. In Sect. 4, we examine the convergence rates when the number of underlying stochastic factors is increased and in Sect. 5, we provide evidence on the convergence rate in the multi-period pricing problem. Next, Sect. 6 provides results for two extensions which are not covered by the existing theory. Finally, Sect. 7 contains concluding remarks. In Appendix we discuss two important numerical issues.

2 American Option Pricing

In this section we introduce a general theoretical framework for American option pricing, we discuss how these options can be priced using simulation and regression, and we review existing theoretical results when it comes to convergence. In the final subsection, we report some preliminary numerical results.

2.1 *Theoretical Framework*

The theoretical framework we use follows closely that of [22, 23], though without any loss of generality we exclude explicit discounting in this presentation. Thus, we first of all assume that the time to expiration can be divided into K periods, $t_0 = 0 < t_1 \leq t_2 \leq \ldots \leq t_K = T$ and we assume a complete probability space $(\Omega, \mathscr{F}, P)$ equipped with a discrete filtration $(\mathscr{F}(t_k))_{k=0}^K$. The underlying model is assumed to be Markovian, with state variables $(X(t_k))_{k=0}^K$ adapted to $(\mathscr{F}(t_k))_{k=0}^K$. We further denote by $(Z(t_k))_{k=0}^K$ an adapted payoff process for the derivative, satisfying $Z(t_k) = h(X(t_k), t_k)$, for a suitable function $h(\cdot, \cdot)$. As an example, consider the American put option for which the only state variable of interest is the stock price, $X(t_k) = S(t_k)$. We have that $Z(t_k) = \max(\bar{S} - S(t_k), 0)$, where $\bar{S}$ denotes the strike price. We assume that $X(0) = x$ is known and hence $Z(0)$ is deterministic. Finally, we let $\mathscr{T}(t_k)$ denote the set of all stopping times with values in $\{t_k, \ldots, t_K\}$.

Following, e.g., [10, 16], in the absence of arbitrage we can specify the object of interest, the American option price, as

$$P(0) = \max_{\tau(t_1) \in \mathscr{T}(t_1)} E[Z(\tau)]. \tag{1}$$

The problem of determining the American option price in (1) is referred to as a discrete time optimal stopping time problem, and one of the preferred ways to solve such problems is to use the dynamic programming principle. Intuitively, this procedure can be motivated by considering the choice faced by the option holder at time t_k, i.e., to exercise the option immediately or to continue to hold the option until the next period. The optimal choice is to exercise immediately if the value of this is positive and larger than the expected payoff from continuing to hold the option.

To fix notation, in the following we let $V(t_k)$ denote the value of the option at a time t_k prior to expiration. If the option holder keeps the option until the next period and acts optimally from this time onwards, the expected payoff will be $E[V(t_{k+1})|X(t_k)]$. That is, the value of continuing to hold the option is the conditional expected option value at time t_{k+1}. On the other hand, if the option is exercised immediately the payoff is $Z(t_k)$. Thus, the value of the option at time t_k may be written as

$$V(t_k) = \max(Z(t_k), E[V(t_{k+1})|X(t_k)]). \tag{2}$$

Furthermore, since the value at expiration equals the intrinsic value, the value functions can be generated iteratively according to the following algorithm:

$$\begin{cases} V(t_K) = Z(t_K) \\ V(t_k) = \max\left(Z(t_k), E\left[V(t_{k+1})|X(t_k)\right]\right), \ k \leq K-1. \end{cases} \tag{3}$$

From the algorithm in (3), the value of the option in (1) can be calculated as

$$P(0) = E\left[V(t_1)|X(0)\right]. \tag{4}$$

The backward induction theorem of [8] (Theorem 3.2) provides the theoretical foundation for the algorithm in (3) and establishes the optimality of the price in (4).

Alternatively, one can formulate the dynamic programming problem directly in terms of the stopping times. To illustrate this, we denote by $C(\tau(t_k)) = Z(\tau(t_k))$ the cash flow generated by the option from following the stopping strategy $\tau(t_k)$ from t_k to expiration. Since the choice facing the option holder at any time t_k prior to expiration is whether to exercise or to continue to hold the option until the next period, it follows that the price of the option at time t_k may be written as

$$V(t_k) = \max\left(Z(t_k), E\left[C(\tau(t_{k+1}))|X(t_k)\right]\right), \tag{5}$$

where $E\left[C(\tau(t_{k+1}))|X(t_k)\right]$ denotes the expected future cash flow. Furthermore, since it is always optimal to exercise at expiration, the optimal stopping time $\tau(t_k)$ can be generated iteratively according to the following algorithm:

$$\begin{cases} \tau(t_K) = T \\ \tau(t_k) = t_k 1_{\{Z(t_k) \geq E[C(\tau(t_{k+1}))|X(t_k)]\}} + \tau(t_{k+1}) 1_{\{Z(t_k) < E[C(\tau(t_{k+1}))|X(t_k)]\}}, \ k \leq K-1, \end{cases} \tag{6}$$

From the algorithm in (6), the value of the option in (1) can be calculated as

$$P(0) = E\left[C(\tau(0))|X(0)\right]. \tag{7}$$

By definition, this price corresponds to what is obtained with (4) and hence share the same optimality properties.

2.2 *American Option Pricing Using Simulation and Regression*

The main problem with the two methods outlined above is that in general the conditional expectation functions $E\left[V(t_{k+1})|X(t_k)\right]$ and $E\left[C(\tau(t_{k+1}))|X(t_k)\right]$ are unknown. In the literature, several methods have been proposed to approximate these conditional expectations using Monte Carlo simulation. Examples include the bundling or state space partitioning methods of [2, 25], the simulated tree method of [5], and the stochastic mesh method of [6]. By now though, the most important methods are probably the regression-based methods of, e.g., [7, 17, 26] in which options are priced by combining the simulation method with cross-sectional regression techniques. In particular, working backwards from the expiration point, at any point in time where exercise should be considered these methods estimate the conditional expectations needed in (3) and (6) by regressing the future values of cash flows on functions of the current state variables. With these estimated conditional expectations, an estimated optimal early exercise strategy can be derived and the option can be priced.

We now consider how to implement this method in the case of the value function algorithm in (3). For notational convenience we define the function $H(\omega, t_k) \equiv E[V(\omega, t_{k+1}) | X(\omega, t_k)]$, where ω represents a sample path. However, the conditional expectation function $H(\omega, t_k)$ is generally unknown and hence some approximation scheme is needed. We let $H_M(\omega, t_k)$ denote an approximation to $H(\omega, t_k)$ based on M terms such that $H_M(\omega, t_k) = \sum_{m=1}^{M} \phi_m(X(\omega, t_k)) b_m(t_k)$, where $\{\phi_m(\cdot)\}_{m=1}^{M}$ form a basis. Furthermore, we assume that some estimation procedure exists, and this is were regression will be used, and that, based on N observations, an approximation, $\tilde{H}_M^N(\omega, t_k)$, to $H_M(\omega, t_k)$ can be constructed as $\tilde{H}_M^N(\omega, t_k) = \sum_{m=1}^{M} \phi_m(X(\omega, t_k)) \tilde{b}_m^N(t_k)$. An estimated approximation to the value function algorithm in (3) can then be derived from the following algorithm:

$$\begin{cases} \tilde{V}_M^N(\omega, t_K) = Z(\omega, t_K) \\ \tilde{V}_M^N(\omega, t_k) = \max\left(Z(\omega, t_k), \tilde{H}_M^N(\omega, t_k)\right), \; k \le K - 1. \end{cases} \tag{8}$$

Furthermore, if the N observations are independent, a natural estimate of the option value in (4) can be calculated as

$$\tilde{P}_M^N(0) = \frac{1}{N} \sum_{n=1}^{N} \tilde{V}_M^N(\omega_n, t_1), \tag{9}$$

since at $t_k = 0$, we have $X(\omega, t_k) = x$. In [25], the approximations used for $\tilde{H}_M^N(\omega, t_k)$ are estimated using splines or a local polynomial smoother. However, other approximation schemes could equally well used and in this paper we will use simple least squares regression.

For the stopping time algorithm in (6), a similar approach can be taken. Again we define the function $F(\omega, t_k) \equiv E[C(\omega, \tau(\omega, t_{k+1})) | X(\omega, t_k)]$, and we let $F_M(\omega, t_k)$ denote an approximation to $F(\omega, t_k)$ based on M terms such that $F_M(\omega, t_k) = \sum_{m=1}^{M} \phi_m(X(\omega, t_k)) a_m(t_k)$. Furthermore, we assume that an approximation, $\hat{F}_M^N(\omega, t_k)$, to $F_M(\omega, t_k)$ can be constructed as $\hat{F}_M^N(\omega, t_k) = \sum_{m=1}^{M} \phi_m(X(\omega, t_k)) \hat{a}_m^N(t_k)$. An estimated approximation to the optimal stopping time algorithm in (6) can then be derived from the following algorithm:

$$\begin{cases} \hat{\tau}_M^N(\omega, t_K) = T \\ \hat{\tau}_M^N(\omega, t_k) = t_k 1_{\{Z(\omega, t_k) \ge \hat{F}_M^N(\omega, t_k)\}} + \hat{\tau}_M^N(\omega, t_{k+1}) 1_{\{Z(\omega, t_k) < \hat{F}_M^N(\omega, t_k)\}}, \; k \le K - 1. \end{cases} \tag{10}$$

Furthermore, if the N observations are independent a natural estimate of the option value in (7) can be calculated as

$$\hat{P}_M^N(0) = \frac{1}{N} \sum_{n=1}^{N} C\left(\omega_n, \hat{\tau}_M^N(\omega_n, 0)\right), \tag{11}$$

since at $t_k = 0$, we have $X(\omega, t_k) = x$. In [17], the approximations used for $\hat{F}_M^N(\omega, t_k)$ are estimated using simple least squares regression.

We end this section by noting that though the value function and stopping time iteration methods are theoretically equivalent, in actual applications they often lead to different results. Moreover, the price estimates obtained with the two estimated approximate algorithms in (8) and (10) may have very different properties. In particular, as shown in [23] the stopping time algorithm has better properties and leads to less biased results particularly when considering the situation with multiple early exercises. The reason is that there is less accumulation of errors when the actual future payoffs from the estimated approximate stopping times are used instead of the estimated approximate value functions in the cross-sectional regressions (see [23] for more on this and for some numerical results).

2.3 Review of Existing Theoretical Results

From the description above of the two methods, it is clear that a central part of the algorithm is the approximation of the conditional expectation functions. Moreover, it is immediately clear that if $\tilde{H}_M^N(\omega, t_k)$ converges to $H(\omega, t_k)$ in the value function iteration algorithm, then the estimated option price in (9) also converges to the true price. For the estimated approximate stopping time algorithm, this may be less obvious. However, one can show that if the conditional expectation approximations converge to the true conditional expectations, then the estimated option price in (11) also converges to the true price. The following proposition, the proof of which can be found in [22], states this formally:

Proposition 1. *If $\hat{a}_M^N(0)$ converges to $a(0)$ as N tends to infinity, then $\hat{P}_M^N(0)$ converges to $P(0)$ in probability.*

Proposition 1 allows us to focus attention on convergence of the conditional expectation approximations when examining the convergence of the algorithm.

We note that it is immediately clear from the definition that when $M \to \infty$, the approximations in $F_M(\omega, t_k)$ converges to $F(\omega, t_k)$ and those in $H_M(\omega, t_k)$ converges to $H(\omega, t_k)$ (see, e.g., [17, 22] for the exact conditions). Moreover, in a Monte Carlo simulation-based setup, it follows that under very general conditions $\hat{F}_M^N(\omega, t_k)$ converges to $F_M(\omega, t_k)$ and $\tilde{H}_M^N(\omega, t_k)$ converges to $H_M(\omega, t_k)$ when $N \to \infty$ (see, e.g., [9] for results for the stopping time algorithm and [26] for results for the value function algorithm). However, it is clear that in order to obtain convergence of $\hat{F}_M^N(\omega, t_k)$ to $F(\omega, t_k)$ and $\tilde{H}_M^N(\omega, t_k)$ to $H(\omega, t_k)$, respectively, both M and N should tend to infinity.

2.3.1 Convergence

The first paper to examine convergence as both M and N tend to infinity in the LSM method is to our knowledge [22]. In particular, Theorem 2 of that paper proves convergence in the general multi-period setting and provides the mathematical foundation for the use of the LSM method in derivatives research. Below we cite this theorem as Theorem 1:

Theorem 1 (Theorem 2 of [22]). *If $M = M(N)$ is increasing in N such that $M \to \infty$ and $M^3/N \to 0$, then $\hat{F}_M^N(\omega, t_k)$ converges to $F(\omega, t_k)$ in probability, for $k = 1, \ldots, K$.*[2]

Together with Proposition 1, this theorem shows that the LSM price estimate converges to the true price under certain regularity assumptions when N tends to infinity provided that M as a function of N tends to infinity as well and that this is in such proportions that $M^3/N \to 0$. In [22], it is noted that this rate is close to the optimal rate in terms of how fast M is allowed to increase, a rate which in the literature on nonparametric estimation is often found to be between M^3 and M^5 (see also [19]). Note that Theorem 1 clearly shows that both the number of paths and the number of regressors should tend to infinity to obtain convergence of the American option price estimate from the LSM method.

Theorem 1 essentially shows that the LSM method converges as long as the number of regressors is not increased too fast, i.e., as long as $M < C \times N^{1/3}$. One of the important assumptions required to obtain this is that the support of the underlying is bounded. In [14], convergence is studied in the unbounded case. This situation complicates the analysis and [14] limit their attention to the normal and log-normal case. They prove convergence as long as $M < C \times \ln(N)$ in the normal case and $M < C \times \sqrt{\ln(N)}$ in the log-normal case. Thus, in the unbounded case the speed with which M can be increased is much slower than it is in the bounded case (see [12] for generalizations of these results to other processes). The intuition for this result is that in the unbounded situation, M has to be increased at a slower rate to avoid problems with the second moment matrix becoming singular.

2.3.2 Convergence Rates

In some situations, it is possible to go further than just proving convergence and to obtain actual convergence rates. Such rates are of interest for several reasons. First of all, by maximizing the rates the fastest possible convergence is obtained. Secondly, once the convergence rates are obtained improved estimates may be obtained using extrapolation techniques. The first paper to derive convergence rates for the LSM method is to our knowledge [22]. In particular, Theorem 1 of that paper shows that under certain conditions in a two-period setting, the conditional expectation

---

[2]See [22] for the proof and the necessary assumptions.

approximation is mean square convergent with known convergence rates. Below we cite this theorem as Theorem 2:

Theorem 2 (Theorem 1 of [22]). *If $M = M(N)$ is increasing in N such that $M \to \infty$ and $M^3/N \to 0$, then the power series estimator $\hat{F}_M^N(\omega)$ is mean square convergent with*

$$\int \left[F(\omega) - \hat{F}_M^N(\omega) \right]^2 dF_0(x) = O_p\left(M/N + M^{-2s/r} \right), \tag{12}$$

where $F_0(x)$ denotes the cumulative distribution function of x, s is the number of continuous derivatives of the conditional expectation function that exist, and r is the dimension of x.[3]

Based on the above theorem one can derive the optimal rate, i.e., the fastest possible convergence rate for the LSM method in the two-period setting. This is obtained by equating the two terms in (12) and obtains when M is proportional to $N^{r/(r+2s)}$. The optimal rate is then given by $N^{-2s/(r+2s)}$. Thus, for options with very smooth payoffs Theorem 2 shows that the convergence rate can be made arbitrarily close to N^{-1}. Moreover, in this case the number of paths, N, should be increased much faster than the number of regressors, M, since $r/(r+2s)$ will be very close to zero. Note that this optimal rate still ensures that $M^3/N \to 0$ as long as $r < s$, i.e., that the number of continuous derivatives is larger than the dimension of x.

More generally, in a multi-period setting convergence rates are difficult to obtain due to the pathwise dependence between future payoffs introduced by the cross-sectional regressions. However, it is possible to derive convergence rates in a modified version of the LSM method where a new set of independently simulated paths is used for the regressions at each time step. The reason for this is that when a new set of simulated paths is used at each step, the future cash flows are independent and this allows us to use Theorem 2 at each possible exercise time. However, in this case the obtained convergence rates are in terms of $K \times N$ instead of N. While this is clearly of interest, the use of additional samples increases the computational time. Thus, obtaining such rates or even estimates of them in the multi-period setting with the "regular" LSM method is clearly of importance.

2.3.3 Results for the Estimated Price

Though [14] consider convergence of the coefficient vectors, in the literature convergence is most often analyzed in terms of the estimated conditional expectation in $\hat{F}_M^N(\omega)$ as it is done above. However, the object of interest is oftentimes the actual price estimate and, though Proposition 1 shows that convergence of the conditional expectations ensures convergence of the estimated price, it remains an open question how fast the price estimate actually converges. Again, the reason it is difficult to obtain such rates is that the dependency introduced through the cross-sectional

---

[3]See [22] for the proof and the necessary assumptions.

regression complicates the analysis. In particular, for the estimated price this dependency is present even in the two-period setting and therefore convergence rates are not readily available for this estimate.

It is possible, however, to obtain convergence rates for the price estimate in one particular situation where a new sample of paths is used for pricing. In particular, in this situation, where the price calculations are done using a different set of simulated paths than that used to determine the conditional expectation approximations, the convergence rate of the American option price estimate is equal to N^{-1} (see also [22]). This is the optimal rate one could hope for and the convergence rate of the corresponding simulated European price estimate. Again, while this is clearly of interest, the use of additional samples increases the computational time. Thus, obtaining such rates or even estimates of them for the estimated price with the "regular" LSM method is clearly also of importance.

2.4 Some Preliminary Numerical Results

From (10) and the resulting option price formula in (11), it is obvious that any estimate as well as the properties of this estimate will depend on the number of regressors used in the cross-sectional regressions, M, as well as the number of paths used in the simulations, N. In this section, we report some numerical results for the estimated conditional expectations and for the estimated option price as the number of regressors, M, and the number of paths, N, are increased. To be specific, we consider $M = 1, 2, \ldots, 10$ and $N = 10,000, 20,000, \ldots, 100,000$. We use a number of artificial put options, the characteristics of which can be found in the first three columns in Table 1 and in the notes to the table. These options range from being deep in the money to being deep out of the money and two different levels of the volatility are considered. Specifically, these options correspond to the one-year options from [17]. To keep it simple, for the time being we consider one possible exercise date prior to expiration only and we consider the Black-Scholes-Merton framework with constant volatility.

2.4.1 Conditional Expectations

We first consider the conditional expectation estimates and compare these to the true conditional expectation, which in the two period Black-Scholes-Merton world is given by the one-period Black-Scholes-Merton price, $BS(\omega, 1)$. Thus, for each choice of M and N we compare the resulting estimated conditional expectation, $\hat{F}_M^N(\omega, 1)$, to the true conditional expectation, $F(\omega, 1) = BS(\omega, 1)$. To this end, we calculate the sample bias given by

$$\widehat{\mathrm{BIAS}}_{\mathrm{EXP}} = \frac{1}{\tilde{N}} \sum_{n=1}^{\tilde{N}} \left(\hat{F}_M^N(\omega_n, 1) - F(\omega_n, 1) \right), \tag{13}$$

Table 1 Regression results for the conditional expectation estimates

Option			Bias² regression			Variance regression			MSE regression		
Nr	S	σ	$\beta_{BIAS^2,M}$	$\beta_{BIAS^2,N}$	R^2	$\beta_{VAR,M}$	$\beta_{VAR,N}$	R^2	$\beta_{MSE,M}$	$\beta_{MSE,N}$	R^2
1	36	0.2	0.000 (0.0019)	-1.017 (0.0019)	1.000	-3.153 (0.2005)	-0.702 (0.2005)	0.728	-3.080 (0.1926)	-0.715 (0.1926)	0.735
2	36	0.4	0.000 (0.0022)	-0.995 (0.0022)	1.000	-3.046 (0.2076)	-0.713 (0.2076)	0.701	-2.969 (0.1989)	-0.728 (0.1989)	0.709
3	38	0.2	0.000 (0.0033)	-0.983 (0.0033)	0.999	-2.889 (0.1993)	-0.722 (0.1993)	0.697	-2.825 (0.1914)	-0.730 (0.1914)	0.705
4	38	0.4	0.000 (0.0036)	-0.961 (0.0036)	0.999	-2.910 (0.2075)	-0.731 (0.2075)	0.683	-2.836 (0.1983)	-0.738 (0.1983)	0.693
5	40	0.2	0.000 (0.0021)	-1.006 (0.0021)	1.000	-2.591 (0.1973)	-0.739 (0.1973)	0.658	-2.535 (0.1892)	-0.751 (0.1892)	0.668
6	40	0.4	0.000 (0.0030)	-0.967 (0.0030)	0.999	-2.771 (0.2070)	-0.741 (0.2070)	0.664	-2.703 (0.1973)	-0.750 (0.1973)	0.676
7	42	0.2	0.000 (0.0019)	-1.021 (0.0019)	1.000	-2.268 (0.1940)	-0.768 (0.1940)	0.611	-2.221 (0.1848)	-0.780 (0.1848)	0.626
8	42	0.4	0.000 (0.0026)	-0.989 (0.0026)	0.999	-2.635 (0.2060)	-0.749 (0.2060)	0.646	-2.571 (0.1957)	-0.761 (0.1957)	0.659
9	44	0.2	0.000 (0.0038)	-1.018 (0.0038)	0.999	-1.930 (0.1878)	-0.787 (0.1878)	0.559	-1.894 (0.1778)	-0.797 (0.1778)	0.579
10	44	0.4	0.000 (0.0022)	-0.996 (0.0022)	1.000	-2.488 (0.2043)	-0.765 (0.2043)	0.626	-2.428 (0.1932)	-0.776 (0.1932)	0.642

Notes: This table reports the results from the regressions in (16)–(18) for the conditional expectations estimates. Standard errors are shown in parentheses. Early exercise is possible only at one time prior to expiration at which time the conditional expectations are estimated. The second and third columns show the stock level, S, and the volatility, σ. All options have a strike price of 40, one year to expiration, and an interest rate of 6% is used.

where $\tilde{N}$ is the number of in-the-money paths. We also calculate the sample variance given by

$$\widehat{\text{VAR}}_{\text{EXP}} = \frac{1}{\tilde{N}} \sum_{n=1}^{\tilde{N}} \left(\hat{F}_M^N (\omega_n, 1) - \bar{F}_M^N (\omega_n, 1) \right)^2, \tag{14}$$

where $\bar{F}_M^N (\omega_n, 1)$ is the mean of $\hat{F}_M^N (\omega_n, 1)$. Finally, we calculate the sample mean squared error, MSE, given by

$$\widehat{\text{MSE}}_{\text{EXP}} = \frac{1}{\tilde{N}} \sum_{n=1}^{\tilde{N}} \left[\hat{F}_M^N (\omega_n, 1) - F (\omega_n, 1) \right]^2. \tag{15}$$

Alternatively, the MSE can be obtained as $\widehat{\text{MSE}}_{\text{EXP}} = (\widehat{\text{BIAS}}_{\text{EXP}})^2 + \widehat{\text{VAR}}_{\text{EXP}}$.

We repeat the calculations 1,000 times using different seeds to generate the $N \times 2$ random numbers used to calculate the stock price paths in the LSM method (see Appendix for further discussion). The averages of the calculated sample bias, variance, and MSE are denoted $\overline{\text{BIAS}}_{\text{EXP}}$, $\overline{\text{VAR}}_{\text{EXP}}$, and $\overline{\text{MSE}}_{\text{EXP}}$, respectively. Instead of reporting the actual averages, we report the results from estimating the following regressions:

$$\ln \left(\overline{\text{BIAS}}_{\text{EXP}} \right)^2 = \alpha + \beta_{\text{BIAS}^2, M} \ln (M) + \beta_{\text{BIAS}^2, N} \ln (N) + \varepsilon, \tag{16}$$

$$\ln \left(\overline{\text{VAR}}_{\text{EXP}} \right) = \alpha + \beta_{\text{VAR}, M} \ln (M) + \beta_{\text{VAR}, N} \ln (N) + \varepsilon, \text{and} \tag{17}$$

$$\ln \left(\overline{\text{MSE}}_{\text{EXP}} \right) = \alpha + \beta_{\text{MSE}, M} \ln (M) + \beta_{\text{MSE}, N} \ln (N) + \varepsilon. \tag{18}$$

In these regressions, in particular the one in (18), we expect the β's to be negative. The results are shown in Table 1. Note that we have left out the "catch all" constant term α since this is uninteresting in terms of convergence.

In columns four through six in Table 1, we report the results from the bias regression in (16). The results show that the estimated value of $\beta_{\text{BIAS}^2, M}$ is close to zero, whereas it is close to -1 for $\beta_{\text{BIAS}^2, N}$. The implication of this is that the bias of the estimated conditional expectation tends to zero as the number of paths used in the regression is increased. In columns seven through nine, we report the results for the variance regression in (17). The results imply that the variance tends to zero both when the number of regressors and when the number of paths are increased.

Finally, in columns ten through twelve we report the results for the mean squared error regression in (18). The results show that the estimated values of $\beta_{\text{MSE}, M}$ and $\beta_{\text{MSE}, N}$ are both significantly negative. This indicates that increasing both the number of regressors and the number of paths should decrease the MSE of the conditional expectation approximation.

2.4.2 Price Estimates

We now consider the American option price estimates. We first of all calculate the bias of the 1,000 estimates which is given by

$$\overline{\text{BIAS}}_{\text{US}} = \frac{1}{1000} \sum_{i=1}^{1000} \left(\hat{P}_M^{N,i}(0) - P_{\text{BM}} \right), \tag{19}$$

where $\hat{P}_M^{N,i}(0)$ is the ith estimate of $P(0)$ and P_{BM} denotes the estimated price from the Binomial Model with 25,000 discretization points to expiration and one possible early exercise date. Similarly, we calculate the variance of the estimates given by

$$\overline{\text{VAR}}_{\text{US}} = \frac{1}{1000} \sum_{i=1}^{1000} \left(\hat{P}_M^{N,i}(0) - \bar{P}_M^{N,i}(0) \right)^2, \tag{20}$$

where $\bar{P}_M^{N,i}(0)$ is the mean estimated price. Finally, we calculate the mean squared error given by

$$\overline{\text{MSE}}_{\text{US}} = \frac{1}{1000} \sum_{i=1}^{1000} \left(\hat{P}_M^{N,i}(0) - P_{\text{BM}} \right)^2. \tag{21}$$

Similar expressions can be calculated for the European price estimates by using the Black-Scholes-Merton price instead of the Binomial Model price in (19) and (21), and we refer to these numbers with subscript "EU" instead of "US." As was the case for the conditional expectation estimates, instead of reporting the actual averages we report the results from estimating the following regressions:

$$\ln \left(\overline{\text{BIAS}}_{\text{US}} \right)^2 = \alpha + \beta_{\text{BIAS}^2,M} \ln (M) + \beta_{\text{BIAS}^2,N} \ln (N) + \varepsilon, \tag{22}$$

$$\ln \left(\overline{\text{VAR}}_{\text{US}} \right) = \alpha + \beta_{\text{VAR},M} \ln (M) + \beta_{\text{VAR},N} \ln (N) + \varepsilon, \text{ and} \tag{23}$$

$$\ln \left(\overline{\text{MSE}}_{\text{US}} \right) = \alpha + \beta_{\text{MSE},M} \ln (M) + \beta_{\text{MSE},N} \ln (N) + \varepsilon. \tag{24}$$

In these regressions, in particular in (24), we expect the β's to be negative. The results are shown in Table 2 for the American price and in Table 3 for the European price. Note that we have again left out the "catch all" constant term α since this is uninteresting in terms of convergence.

In columns two through four in Table 2, we report the results from the bias regression in (22). Although the results are somewhat mixed with respect to $\beta_{\text{BIAS}^2,N}$ both in terms of significance and in terms of the sign, the table shows that $\beta_{\text{BIAS}^2,M}$ is significantly negative in all but one of the cases. This makes sense since whether to exercise or not depends crucially on the conditional expectation which

Table 2 Regression results for the American option price estimates

	Bias² regression			Variance regression			MSE regression		
Nr	$\beta_{BIAS^2,M}$	$\beta_{BIAS^2,N}$	R^2	$\beta_{VAR,M}$	$\beta_{VAR,N}$	R^2	$\beta_{MSE,M}$	$\beta_{MSE,N}$	R^2
1	−4.324 (0.2599)	1.376 (0.2599)	0.759	−0.009 (0.0034)	−0.985 (0.0034)	0.999	−0.545 (0.0496)	−0.899 (0.0496)	0.823
2	−2.703 (0.3488)	0.344 (0.3488)	0.386	−0.012 (0.0044)	−0.986 (0.0044)	0.998	−0.079 (0.0186)	−0.962 (0.0186)	0.965
3	−3.951 (0.2717)	0.846 (0.2717)	0.695	−0.011 (0.0031)	−0.980 (0.0031)	0.999	−0.216 (0.0169)	−0.933 (0.0169)	0.971
4	−1.855 (0.3427)	−0.304 (0.3427)	0.237	−0.003 (0.0038)	−0.985 (0.0038)	0.999	−0.045 (0.0123)	−0.969 (0.0123)	0.985
5	−3.245 (0.2668)	0.478 (0.2668)	0.609	−0.004 (0.0037)	−0.993 (0.0037)	0.999	−0.054 (0.0088)	−0.976 (0.0088)	0.992
6	−2.150 (0.3378)	−0.299 (0.3378)	0.299	−0.003 (0.0042)	−0.990 (0.0042)	0.998	−0.029 (0.0086)	−0.980 (0.0086)	0.993
7	−1.214 (0.3207)	−1.283 (0.3207)	0.238	−0.002 (0.0031)	−1.003 (0.0031)	0.999	−0.018 (0.0054)	−0.996 (0.0054)	0.997
8	−2.343 (0.3989)	−0.779 (0.3989)	0.283	0.003 (0.0034)	−0.998 (0.0034)	0.999	−0.016 (0.0058)	−0.991 (0.0058)	0.997
9	−0.067 (0.3555)	−2.520 (0.3555)	0.342	0.000 (0.0038)	−1.013 (0.0038)	0.999	−0.006 (0.0044)	−1.011 (0.0044)	0.998
10	−2.845 (0.3672)	−2.063 (0.3672)	0.486	0.007 (0.0026)	−1.001 (0.0026)	0.999	−0.012 (0.0033)	−0.995 (0.0033)	0.999

Notes: This table reports the results from the regressions in (22)–(24) for the American option price estimates. Standard errors are shown in parentheses. See the notes to Table 1 for the characteristics of the options.

Table 3 Regression results for the European option price estimates

Nr	Bias² regression			Variance regression			MSE regression		
	$\beta_{BIAS^2,M}$	$\beta_{BIAS^2,N}$	R^2	$\beta_{VAR,M}$	$\beta_{VAR,N}$	R^2	$\beta_{MSE,M}$	$\beta_{MSE,N}$	R^2
1	0.000 (0.3266)	0.625 (0.3266)	0.036	0.000 (0.0032)	−1.002 (0.0032)	0.999	0.000 (0.0032)	−1.002 (0.0032)	0.999
2	0.000 (0.2763)	0.159 (0.2763)	0.003	0.000 (0.0031)	−0.998 (0.0031)	0.999	0.000 (0.0031)	−0.998 (0.0031)	0.999
3	0.000 (0.1698)	0.737 (0.1698)	0.163	0.000 (0.0030)	−1.007 (0.0030)	0.999	0.000 (0.0030)	−1.007 (0.0030)	0.999
4	0.000 (0.2659)	0.257 (0.2659)	0.010	0.000 (0.0029)	−0.999 (0.0029)	0.999	0.000 (0.0028)	−0.998 (0.0028)	0.999
5	0.000 (0.1856)	0.485 (0.1856)	0.066	0.000 (0.0037)	−1.017 (0.0037)	0.999	0.000 (0.0037)	−1.017 (0.0037)	0.999
6	0.000 (0.2248)	0.107 (0.2248)	0.002	0.000 (0.0028)	−1.003 (0.0028)	0.999	0.000 (0.0028)	−1.003 (0.0028)	0.999
7	0.000 (0.2545)	−0.301 (0.2545)	0.014	0.000 (0.0044)	−1.028 (0.0044)	0.998	0.000 (0.0045)	−1.028 (0.0045)	0.998
8	0.000 (0.1564)	1.008 (0.1564)	0.300	0.000 (0.0030)	−1.008 (0.0030)	0.999	0.000 (0.0030)	−1.008 (0.0030)	0.999
9	0.000 (0.2034)	−0.796 (0.2034)	0.136	0.000 (0.0047)	−1.040 (0.0047)	0.998	0.000 (0.0047)	−1.040 (0.0047)	0.998
10	0.000 (0.1656)	1.452 (0.1656)	0.442	0.000 (0.0034)	−1.012 (0.0034)	0.999	0.000 (0.0034)	−1.012 (0.0034)	0.999

Notes: This table reports the results from the regressions in (22)–(24) for the European option price estimates. Standard errors are shown in parentheses. See the notes to Table 1 for the characteristics of the options.

is approximated increasingly accurately as the number of regressors increases.[4] In columns five through seven, we report the results for the variance regression in (23). The results show that the variance tends to zero when the number of paths is increased. Increasing the number of regressors, on the other hand, has only very little and in many cases only an insignificant effect on the variance of the estimate.

Finally, in columns eight through ten we report the results for the mean squared error regression in (24). The results show that both the number of regressors, M, and the number of paths, N, influence the MSE. In particular, the significantly negative values for $\beta_{\mathrm{MSE},M}$ and $\beta_{\mathrm{MSE},N}$ indicate that as both tend to infinity the MSE decreases. Furthermore, from the table it is clear that while the estimated value for $\beta_{\mathrm{MSE},N}$ remains roughly constant the estimates for $\beta_{\mathrm{MSE},M}$ decrease as the stock price, S, increases. Thus, the results indicate that increasing the number of regressors becomes less and less important as we move towards the out of the money region. Intuitively this makes sense since in this region the early exercise feature of the American option becomes less important.

For the purpose of comparison, Table 3 provides the corresponding results for the European price estimate. From the table the importance of increasing N is clear for both the variance and the MSE. On the other hand, there is no significant effect of increasing M, which plays no role what so ever in determining the price of this type of option. The table also confirms that the estimated European option price is essentially unbiased and the mean squared error is therefore entirely due to the variance of the estimated price.

3   Convergence Rates in a Two-Period Setting

Theorem 2 above provides a convergence result for the conditional expectation approximation in a two-period setting. In particular, the theorem shows that the conditional expectation approximation converges to the true conditional expectation in mean square sense when the number of regressors, M, and the number of paths, N, tend to infinity provided that $M^3/N \to 0$. In this section, we perform a Monte Carlo experiment to illustrate this. To ensure that $M^3/N \to 0$ as required in the theorem we first set $N = 10 \times M^4$ for $M = 3, \ldots, 10$, i.e., we use $N = 810, 2,560, 6,250, 12,960, 24,010, 40,960, 65,610$, and $100,000$.[5] We analyze both the convergence rates of

---

[4]The exception to this is option number 9, which is a deep out of the money option with low volatility. For this option, very few paths are in the money and hence used in the cross-sectional regression, and as a result the price estimate is biased upwards due to overfitting. In fact, when using 10 times the number of paths the estimated value for $\beta_{\mathrm{BIAS}^2,M}$ is significantly negative for this option also.

[5]To be precise, what should be increased is the number of in-the-money paths, $\tilde{N}$, used in the cross-sectional regressions. Unfortunately, it is not possible to control directly this number by the nature of the Monte Carlo study. However, the proportion of the paths that are in the money should be approximately constant and we will therefore have that $\tilde{N} \propto M^4$.

the estimated conditional expectations and the estimated price. However, this initial choice should not be taken as a recommendation of which values to choose for M and N. Indeed, the theoretical result implies that relatively higher values of N should be chosen as long as the conditional expectation function is smooth. In the last subsection, we return to this.

3.1 Convergence Rates for the Estimated Conditional Expectations

For each of the values of N, and hence also implicitly of M, we calculate the mean of the bias, of the variance, and of the mean squared error from the formulas in (13)–(15), respectively. We maintain the assumption of a two-period Black-Scholes-Merton model since this allows us to use the one-period Black-Scholes-Merton European option price as the true conditional expectation. Instead of reporting the actual averages, we report the results of performing the following regressions:

$$\ln\left(\overline{\text{BIAS}}_{\text{EXP}}\right)^2 = \alpha + \beta_{\text{BIAS}^2} \ln(N) + \varepsilon, \tag{25}$$

$$\ln\left(\overline{\text{VAR}}_{\text{EXP}}\right) = \alpha + \beta_{\text{VAR}} \ln(N) + \varepsilon, \text{ and} \tag{26}$$

$$\ln\left(\overline{\text{MSE}}_{\text{EXP}}\right) = \alpha + \beta_{\text{MSE}} \ln(N) + \varepsilon. \tag{27}$$

Compared to the regressions in (16)–(18), the $\ln(M)$ term is eliminated in these regressions due to the linear relationship between $\ln(N)$ and $\ln(M)$.

Panel A of Table 4 provides results on the convergence of the estimated conditional expectations. The first thing to note from the table is the very high values for the R^2 indicating the validity of the relations in (25)–(27). Furthermore, the estimated β's clearly indicate that the estimated conditional expectation converges to the true conditional expectation as these are all significantly negative. The results also show that, in general, the bias converges faster towards zero than the variance as the estimated β_{BIAS^2}'s are numerically larger than the estimated β_{VAR}'s. However, Theorem 2 not only provides the convergence result it also provides theoretical rates of convergence for the estimated conditional expectation. For example, with the present choice of $N \propto M^4$ the theorem reads

$$\int \left[F(\omega) - \hat{F}_M^N(\omega)\right]^2 dF_0(x) = O_p\left(N^{-3/4} + N^{-s/2}\right). \tag{28}$$

We see that the second term in (28) is negligible as long as the conditional expectation function is (at least) twice continuously differentiable. This is obviously the case in the present setting, and hence the theoretical order of convergence is equal to $-3/4$. The estimated values for β_{MSE} are all below -0.75 in value confirming numerically the result of Theorem 2.

Table 4 Convergence results

Panel A: Conditional expectation estimates

Nr.	β_{BIAS^2}			β_{VAR}			β_{MSE}		
	Estim	S.E.	R^2	Estim	S.E.	R^2	Estim	S.E.	R^2
1	−0.989	(0.007)	1.000	−0.714	(0.005)	1.000	−0.768	(0.005)	1.000
2	−0.992	(0.005)	1.000	−0.713	(0.003)	1.000	−0.775	(0.004)	1.000
3	−0.984	(0.007)	1.000	−0.706	(0.004)	1.000	−0.759	(0.002)	1.000
4	−0.982	(0.009)	0.999	−0.710	(0.004)	1.000	−0.770	(0.003)	1.000
5	−0.981	(0.004)	1.000	−0.704	(0.005)	1.000	−0.754	(0.002)	1.000
6	−0.982	(0.008)	1.000	−0.706	(0.006)	1.000	−0.767	(0.002)	1.000
7	−0.969	(0.009)	1.000	−0.707	(0.006)	1.000	−0.752	(0.003)	1.000
8	−0.979	(0.005)	1.000	−0.711	(0.006)	1.000	−0.768	(0.002)	1.000
9	−0.980	(0.006)	1.000	−0.702	(0.005)	1.000	−0.748	(0.003)	1.000
10	−0.970	(0.005)	1.000	−0.717	(0.006)	1.000	−0.769	(0.002)	1.000

Panel B: American option price estimates

Nr.	β_{BIAS^2}			β_{VAR}			β_{MSE}		
	Estim	S.E.	R^2	Estim	S.E.	R^2	Estim	S.E.	R^2
1	−1.523	(0.284)	0.827	−1.002	(0.010)	0.999	−1.004	(0.011)	0.999
2	−2.334	(0.399)	0.851	−1.000	(0.008)	1.000	−1.003	(0.009)	1.000
3	−1.802	(0.275)	0.878	−0.996	(0.008)	1.000	−0.997	(0.008)	1.000
4	−2.277	(0.491)	0.782	−0.998	(0.007)	1.000	−1.000	(0.007)	1.000
5	−2.415	(0.820)	0.591	−0.998	(0.006)	1.000	−1.000	(0.006)	1.000
6	−2.022	(0.246)	0.918	−1.002	(0.007)	1.000	−1.004	(0.007)	1.000
7	−1.971	(0.116)	0.980	−1.004	(0.005)	1.000	−1.005	(0.005)	1.000
8	−2.154	(0.391)	0.835	−1.004	(0.004)	1.000	−1.006	(0.004)	1.000
9	−1.914	(0.145)	0.966	−1.008	(0.006)	1.000	−1.011	(0.006)	1.000
10	−2.135	(0.304)	0.892	−1.005	(0.003)	1.000	−1.007	(0.004)	1.000

Panel C: European option price estimates

Nr.	β_{BIAS^2}			β_{VAR}			β_{MSE}		
	Estim	S.E.	R^2	Estim	S.E.	R^2	Estim	S.E.	R^2
1	−1.409	(0.649)	0.440	−0.998	(0.005)	1.000	−0.998	(0.005)	1.000
2	−1.450	(0.763)	0.375	−0.998	(0.005)	1.000	−0.999	(0.005)	1.000
3	−1.139	(0.842)	0.234	−0.998	(0.004)	1.000	−0.998	(0.004)	1.000
4	−1.391	(0.776)	0.349	−0.998	(0.005)	1.000	−0.998	(0.005)	1.000
5	−1.255	(0.557)	0.459	−0.999	(0.004)	1.000	−0.999	(0.004)	1.000
6	−1.403	(0.814)	0.331	−0.998	(0.004)	1.000	−0.998	(0.004)	1.000
7	−1.147	(0.632)	0.355	−0.999	(0.007)	1.000	−0.999	(0.007)	1.000
8	−1.104	(0.412)	0.545	−0.999	(0.004)	1.000	−0.999	(0.004)	1.000
9	−1.456	(0.415)	0.672	−1.001	(0.009)	1.000	−1.001	(0.009)	1.000
10	−1.140	(0.373)	0.609	−0.999	(0.004)	1.000	−0.999	(0.004)	1.000

Notes: This table reports the convergence results. Panel A reports the results from the regressions in (25)–(27) for the conditional expectation estimates. Panel B reports the results from the regressions in (29)–(31) for the American option price estimates and Panel C reports the results for the European option price estimate. See the notes to Table 1 for the characteristics of the options.

3.2 Convergence Rates for the Estimated Price

Next, we calculate the bias, variance, and mean squared error for the estimated American and European option prices using the formulas in (19)–(21), respectively. Again, instead of reporting the actual averages we report the results of the following regressions:

$$\ln\left(\overline{\text{BIAS}}_{\text{US}}\right)^2 = \alpha + \beta_{\text{BIAS}^2}\ln(N) + \varepsilon, \qquad (29)$$

$$\ln\left(\overline{\text{VAR}}_{\text{US}}\right) = \alpha + \beta_{\text{VAR}}\ln(N) + \varepsilon, \text{ and} \qquad (30)$$

$$\ln\left(\overline{\text{MSE}}_{\text{US}}\right) = \alpha + \beta_{\text{MSE}}\ln(N) + \varepsilon. \qquad (31)$$

Compared to the regressions in (22)–(24), the $\ln(M)$ term is eliminated in these regressions due to the linear relationship between $\ln(N)$ and $\ln(M)$.

Panel B of Table 4 provides results on the convergence of the estimated American option prices. From Theorem 2 and Proposition 1 above we know that the American price estimate converges to the true value although the rate is unknown. The first thing to note from the table is that the convergence is confirmed by the estimated values of β_{BIAS^2}, β_{VAR}, and β_{MSE} which are all significantly negative. This indicates that both the bias and the variance, and hence also the mean squared error, tend to zero as the number of paths tends to infinity. Note also that the estimated β_{BIAS^2}'s are very large numerically, and as a result the estimated β_{MSE}'s are very close to -1, the estimate obtained for the variance term. Thus, the numerical results indicate that the convergence rate of the estimated American option price is in fact close to the optimal rate of N^{-1}.

For the purpose of comparison, Panel C of Table 4 provides the corresponding results for the estimated European option prices. For the estimated European prices, theoretical results exist proving that the price estimate is unbiased and has a convergence rate of N^{-1}. This is confirmed by the results in the table which show that the estimated β_{MSE}'s are insignificantly different from -1. Moreover, the table shows that for the bias the R^2 values are quite low and the standard errors on the estimated β_{BIAS^2}'s very large. As the European price estimates should be unbiased, this is to be expected.

3.3 The Effect of Increasing N Relatively Faster

As mentioned in the beginning of this section, setting $N \propto M^4$ is not necessarily optimal. Indeed, the theoretical results imply that N should be increased relatively faster as long as the conditional expectation function is smooth. In fact, the optimal choice is to set $M \propto N^{r/(r+2s)}$, and with this choice the convergence rate obtained is given by $N^{-2s/(r+2s)}$. In this section we examine this issue by changing the coefficient of proportionality, γ, in $N = C \times M^\gamma$ from $\gamma = 1$ to $\gamma = 10$ for $M =$

$5, \ldots, 10$. In all cases, we pick C such that for $M = 10$ the resulting $N = 100{,}000$. We note that this choice does not affect the results which are similar when a fixed C is used.

Panel A of Table 5 provides results on the convergence of the estimated conditional expectations for option number 5, the at-the-money option with a volatility of 20%. The first thing to note from the table is that the theoretical bounds on the order of convergence for $\gamma \geq 4$ are confirmed numerically as the estimated β_{MSE}'s are larger in absolute value than the theoretical values given by Theorem 2. The table also shows that the estimated values are very close to the bounds even when $\gamma \leq 3$, a case which is not covered by this theorem and for which the bounds may not hold true. When considering the results for the bias and variance terms, it is seen that the decrease in the estimated order for low values of γ is driven by the variance term. In fact, for sufficiently low γ, i.e., $\gamma = 1$, this term appears to explode numerically. The bias term, on the other hand, converges fast irrespectively of the choice of γ, a result driven by the smoothness of the conditional expectation function for this type of options.

Panel B of Table 5 provides results for the convergence of the estimated American option prices for option number 5. The table shows that the estimated order of convergence does not change significantly with γ for $\gamma \geq 4$. In particular, as long as $\gamma \geq 4$ the estimate of β_{MSE} is insignificantly different from -1. For values of $\gamma \leq 3$, however, the estimated order of convergence does in fact decrease significantly, though numerically they remain quite high. In fact, this is so even when $\gamma = 1$ and in this case the price still converges though the conditional expectation does not.

4 Multiple Number of Stochastic Factors

The convergence theorem holds more generally than the simple Black-Scholes-Merton case. In particular, Theorem 2 above holds for options on multiple underlying stochastic factors. In this section we examine what happens when the number of stochastic factors, i.e., the dimension of x, is increased. Conceptually, the simplest example of multiple stochastic factors in the option pricing literature is probably the case in which there is more than one underlying asset. Indeed, many of the papers extending the Binomial Model to allow for several underlying state variables have used the case of several underlying assets as a motivating example (see [3,4]).

Thus, in order to examine the issue of convergence and convergence rates with multiple stochastic factors we first of all consider American style put options written on multiple underlying assets with payoff functions given by either

$$\text{Geometric average options}: \ Z(t_k) = \max\left[0, \bar{S} - \left(\prod_{l=1}^{r} S_l(t_k)\right)^{\frac{1}{r}}\right], \ \text{or} \quad (32)$$

Table 5 Convergence results for different choices of γ

Panel A: Conditional expectation estimates

γ	Bound	β_{BIAS^2}			β_{VAR}			β_{MSE}		
		Estim	S.E.	R^2	Estim	S.E.	R^2	Estim	S.E.	R^2
1	0.00	−1.031	(0.032)	0.996	0.119	(0.013)	0.955	−0.042	(0.008)	0.880
2	−0.50	−1.009	(0.017)	0.999	−0.430	(0.006)	0.999	−0.512	(0.007)	0.999
3	−0.67	−0.996	(0.011)	0.999	−0.619	(0.006)	1.000	−0.672	(0.005)	1.000
4	−0.75	−0.988	(0.008)	1.000	−0.717	(0.003)	1.000	−0.754	(0.003)	1.000
5	−0.80	−0.977	(0.014)	0.999	−0.779	(0.003)	1.000	−0.806	(0.004)	1.000
6	−0.83	−0.977	(0.006)	1.000	−0.811	(0.003)	1.000	−0.833	(0.002)	1.000
7	−0.86	−0.983	(0.010)	1.000	−0.842	(0.003)	1.000	−0.861	(0.003)	1.000
8	−0.88	−0.987	(0.005)	1.000	−0.864	(0.001)	1.000	−0.881	(0.002)	1.000
9	−0.89	−0.989	(0.007)	1.000	−0.877	(0.002)	1.000	−0.893	(0.001)	1.000
10	−0.90	−1.002	(0.008)	1.000	−0.889	(0.003)	1.000	−0.905	(0.002)	1.000

Panel B: American option price estimates

γ	β_{BIAS^2}			β_{VAR}			β_{MSE}		
	Estim	S.E.	R^2	Estim	S.E.	R^2	Estim	S.E.	R^2
1	−4.081	(2.316)	0.437	−0.910	(0.008)	1.000	−0.911	(0.008)	1.000
2	−0.895	(1.323)	0.103	−0.943	(0.007)	1.000	−0.943	(0.007)	1.000
3	−0.665	(0.846)	0.134	−0.976	(0.023)	0.998	−0.976	(0.023)	0.998
4	−1.929	(1.867)	0.211	−0.992	(0.009)	1.000	−0.992	(0.010)	1.000
5	−1.467	(0.596)	0.602	−1.018	(0.013)	0.999	−1.018	(0.014)	0.999
6	−2.147	(0.385)	0.886	−1.008	(0.010)	1.000	−1.010	(0.010)	1.000
7	−2.264	(0.385)	0.896	−1.000	(0.009)	1.000	−1.002	(0.009)	1.000
8	−2.153	(0.108)	0.990	−0.990	(0.010)	1.000	−0.994	(0.009)	1.000
9	−2.231	(0.049)	0.998	−0.998	(0.008)	1.000	−1.006	(0.008)	1.000
10	−2.186	(0.286)	0.936	−1.008	(0.008)	1.000	−1.019	(0.010)	1.000

Notes: This table reports the convergence results with $N \propto M^\gamma$, for different choices of M^γ. Panel A reports the results from the regressions in (25)–(27) for the conditional expectation estimates. Panel B reports the results from the regression in (29)–(31) for the American option price estimates. Results are presented for option number 5 (see the notes to Table 1 for the characteristics of the option).

$$\text{Arithmetic average options}: \ Z(t_k) = \max\left[0, \bar{S} - \frac{1}{r}\sum_{l=1}^{r} S_l(t_k)\right]. \tag{33}$$

These payoff functions are smooth in the in the money region, and Theorem 2 should hold true. However, one might also consider options with payoff functions specified as

$$\text{Maximum options}: \ Z(t_k) = \max\left[0, \bar{S} - \max(S_1(t_k), \ldots, S_r(t_k))\right], \text{ or} \tag{34}$$

$$\text{Minimum options}: \ Z(t_k) = \max\left[0, \bar{S} - \min(S_1(t_k), \ldots, S_r(t_k))\right]. \tag{35}$$

These payoff functions are not continuously differentiable of any order in the in-the-money region. Hence, the LSM method could potentially fail to converge unless their conditional expectations are sufficiently smooth. For the present we limit attention to $r = 2$, and perform a Monte Carlo experiment along the lines above. To be specific, we price options with the above payoffs and with the following characteristics for the underlying assets: $S_l(0) = 40$ and $\sigma_l = 0.20$, $l = 1$ and 2, with a correlation of $\rho = 0.25$. The options are at the money with a strike price of 40, with one year to maturity, and the interest rate is set equal to 6%.

In the LSM method the regressors are chosen as described in [21] using power series, increasing the maximum order from three to ten. The resulting number of regressors varies from $M = 10$ to 66. Compared to the one-dimensional case where the number of regressors grows linearly, one might fear that the LSM method suffers from the curse of dimensionality. However, we know from [15] that the number of regressors, M, increases polynomially only, in the number of assets for a given maximum order, $\bar{M}$.[6] We choose N proportional to M^4 with the factor of proportionality C set such that for $\bar{M} = 10$ a total of $N = 100,000$ paths are used and repeat the calculations 1,000 times.[7] Though it is possible to calculate the theoretical price for options on the geometric average, this is not the case for the other payoffs. Hence, for consistency we compare the results to the Binomial Model with the specification of [4] for the jump sizes and the jump probabilities. This method is used for the prices and to obtain an estimate of the true conditional expectation function. To be specific, we take $F(\omega_n, 1) = F_{10}(\omega_n, 1)$, where $F_{10}(\omega_n, 1)$ is the projection of the conditional expectations in the Binomial Model onto a constant and the first nine powers of the stock price. Using ten regressors was enough to explain the variation in the conditional expectation and obtain a $R^2 \approx 1$.

Panel A of Table 6 provides results on the convergence of the estimated conditional expectations for options on two underlying assets with the payoff functions in (32)–(35). The first thing to note from the table is that with the smooth payoff functions for the geometric and arithmetic average options the convergence

---

[6]The total number of regressors with a maximum order of at most m in r dimensions is given by $(m+r)!/(m!r!)$ (see also [11]).

[7]To achieve this we set $C = 0.10928$ and round the number of paths.

Table 6 Convergence results for options on two assets

Panel A: Conditional expectation estimates

Payoff	β_{BIAS^2}			β_{VAR}			β_{MSE}		
	Estim	S.E.	R^2	Estim	S.E.	R^2	Estim	S.E.	R^2
GEOM	−1.004	(0.005)	1.000	−0.737	(0.002)	1.000	−0.749	(0.001)	1.000
ARITH	−1.008	(0.005)	1.000	−0.735	(0.003)	1.000	−0.747	(0.002)	1.000
MAX	−0.988	(0.009)	1.000	−0.711	(0.005)	1.000	−0.723	(0.003)	1.000
MIN	−0.320	(0.073)	0.795	−0.514	(0.042)	0.967	−0.482	(0.048)	0.953

Panel B: American option price estimates

Payoff	β_{BIAS^2}			β_{VAR}			β_{MSE}		
	Estim	S.E.	R^2	Estim	S.E.	R^2	Estim	S.E.	R^2
GEOM	−1.470	(0.060)	0.992	−1.013	(0.007)	1.000	−1.042	(0.011)	0.999
ARITH	−2.094	(0.219)	0.948	−1.014	(0.007)	1.000	−1.043	(0.011)	0.999
MAX	−1.299	(0.342)	0.743	−1.017	(0.004)	1.000	−1.030	(0.011)	0.999
MIN	−0.907	(0.094)	0.949	−1.009	(0.005)	1.000	−0.986	(0.017)	0.998

Notes: This table reports the convergence results for options on two assets. Panel A reports the results from the regressions in (25)–(27) for the conditional expectation estimates. Panel B reports the results from the regressions in (29)–(31) for the American option price estimates. The option characteristics are $S_l(0) = 40$ and $\sigma_l = 0.20$, $l = 1$ and 2, and $\rho_{12} = 0.25$. Furthermore, the option has a strike price of 40, one year to expiration, and an interest rate of 6% is used. Early exercise is possible only at one time prior to expiration at which time the conditional expectations are estimated.

rates are numerically confirmed. In particular, for these options the estimated β_{MSE}'s are very close to the theoretical bound given by -0.75. For the maximum option the estimated order of convergence is also very close to -0.75, though statistically lower. However, for the minimum option the estimated order is much lower than -0.75. This indicates that for this type of payoffs the second term in Theorem 2 is binding. In particular, this is clear when considering the estimated value of β_{BIAS^2} since this second term is linked to the bias of the estimator.

Panel B of Table 6 provides results for the convergence of the estimated American option prices for options on two underlying assets with the payoff functions in (32)–(35). The table shows that for options with smooth payoffs the estimated order of convergence of the price is close to -1 as it was the case for the one-dimensional case. This also holds for the maximum option. However, for the minimum option, the one for which the estimated conditional expectations had the slowest estimated order of convergence, this is in fact somewhat lower than -1 in absolute value.

5 Multiple Number of Exercise Times

In the previous sections we considered the situation with only one early exercise time, which allowed us to use Theorem 2 above. In this section, we examine what happens as the number of exercise times is increased. That is, we essentially examine the validity of Theorem 1 above. Note that, though this theorem provides the mathematical foundation for the use of the LSM method in derivatives pricing in that convergence is shown, no convergence rates are reported. The reason for this is that when more than one early exercise point is considered the pathwise independence is lost since the future payoff depends on estimated conditional expectations. Thus, at any time t in which the dependent variable is calculated from an approximation which has been estimated by a cross-sectional regression, dependence is introduced between the paths.

In order to examine the issue of convergence and convergence rates with multiple early exercises we again choose to price option number 5, the at-the-money option with a volatility of 20%. However, we now consider options with 2, 5, 10, and 25 exercise times. As before, we increase the number of regressors from three to ten in the LSM method, and we set $M = 10 \times K^4$. Again, the calculations are repeated 1,000 times. The results are compared to the Binomial Model with the same number of possible early exercise times. The benchmark value for the conditional expectations is again $F(\omega_n, 1) = F_{10}(\omega_n, 1)$, where $F_{10}(\omega_n, 1)$ is the projection of the conditional expected value in the Binomial Model onto a constant and the first nine powers of the stock price at the earliest possible exercise time. Using ten regressors was enough to explain the variation in the conditional expectation and obtain a $R^2 \approx 1$.[8]

---

[8] Also note that the estimated order of convergence in Table 7 are very close to those in Table 4 in which the true conditional expectation is used.

Table 7 Convergence results for options with multiple early exercises

Panel A: Conditional expectation estimates

	β_{BIAS^2}			β_{VAR}			β_{MSE}		
Steps	Estim	S.E.	R^2	Estim	S.E.	R^2	Estim	S.E.	R^2
2	−0.981	(0.004)	1.000	−0.703	(0.005)	1.000	−0.753	(0.002)	1.000
5	−1.001	(0.011)	0.999	−0.708	(0.005)	1.000	−0.766	(0.004)	1.000
10	−1.021	(0.010)	0.999	−0.741	(0.003)	1.000	−0.797	(0.007)	1.000
25	−1.044	(0.014)	0.999	−0.710	(0.005)	1.000	−0.778	(0.010)	0.999

Panel B: American option price estimates

	β_{BIAS^2}			β_{VAR}			β_{MSE}		
Steps	Estim	S.E.	R^2	Estim	S.E.	R^2	Estim	S.E.	R^2
2	−2.415	(0.820)	0.591	−0.998	(0.006)	1.000	−1.000	(0.006)	1.000
5	−2.884	(0.360)	0.915	−1.002	(0.008)	1.000	−1.011	(0.008)	1.000
10	−1.762	(0.067)	0.992	−1.018	(0.007)	1.000	−1.039	(0.009)	1.000
25	−1.517	(0.031)	0.997	−1.016	(0.008)	1.000	−1.054	(0.010)	0.999

Notes: This table reports the convergence results for options with multiple early exercises. Panel A reports the results from the regressions in (25)–(27) for the conditional expectation estimates at the earliest possible exercise time. Panel B reports the results from the regressions in (29)–(31) for the American option price estimates. Results are presented for option number 5 (see the notes to Table 1 for the characteristics of the option).

Panel A of Table 7 provides results on the convergence of the estimated conditional expectations for option number 5 with increasing number of exercise times. The first thing to note from the table is that the estimated conditional expectation does indeed seem to converge irrespective of the number of steps. In particular, both the bias and the variance tend to zero as the number of paths is increased as indicated by the significantly negative estimated values of β_{BIAS^2} and β_{VAR}. The estimated values for the mean squared error are also significantly negative. In fact, these estimates remain very large numerically and even with 25 exercise times the estimate is above -0.75 in absolute value, the theoretical order of convergence in the two-period case or when independent paths are used at each step.

Panel B of Table 7 provides results for the convergence of the estimated American option prices for option number 5 with increasing number of exercise times. The table shows that for the price estimates no matter the number of early exercise times the bias and the variance of the estimated price tends to zero as the number of paths tends to infinity. Moreover, even with 25 steps the estimated order of convergence is very close to -1. These results not only lend support to the general convergence theorem, they also provide numerical evidence on the excellent performance of the LSM method in the multi-period setting.

6 Other Extensions

Though the analysis above has covered several of the most important cases, there are at least two possibly interesting extensions. The first of these is to analyze the effect of using different polynomial families as regressors. The second is to examine the results when using all the paths in the regression instead of using only the in-the-money paths. In this final section, we provide numerical evidence on the convergence and the convergence rates in these two cases which are not covered by the above theorems.

6.1 Other Regressors: Alternative Orthogonal Polynomial Families

It is important to stress that neither Theorem 1 nor Theorem 2 provides any guidelines as to which set of regressors should be used to approximate the conditional expectation. In particular, the choice of powers as regressors is not essential for the proofs of these two theorems, and they will hold true for any choice of polynomial family $\{\phi_k\}_{m=0}^{M}$ for which there exists a nonsingular constant matrix, B, such that $P_M = B\phi_M$. This covers many well-known families of orthogonal polynomials. In particular, the Laguerre family, the Hermite family as well as the Chebyshev family of both first and second kind fall within the framework used here, since all of these families can be written as ordinary polynomials. To see this note that polynomials from all of the above-mentioned families satisfies the following recursive formula from [1]

$$a_{1k}f_{k+1}(x) = (a_{2k} + a_{3k}x) f_k(x) - a_{4k}f_{k-1}(x),\qquad(36)$$

where the coefficients can be seen in Table 8. For the Legendre polynomials this is exactly the formula which was used in [22], and in terms of the fitted values in the cross-sectional regressions the result would be the same no matter what family is used.

However, the different families mentioned above have different orthogonality properties both in terms of orthogonal interval and in terms of weighting functions as the table shows. For instance, the family of Laguerre polynomials are orthogonal on the interval $(0;\infty)$ with respect to the weighting function $w(x) = \exp(-x)$. The weighted Laguerre polynomials were used by [17] in the cross-sectional regressions to price American put options. However, for this family of polynomials a large degree of collinearity exists between the regressors when the interval of interest is say $(0;1)$ only and this could lead to numerical problems (see, e.g., [18, 21]). Theoretically though, a much more important problem with the weighted Laguerre polynomials is that Theorems 1 and 2 are invalidated by the nonconstant term in the weighting function. In particular, this means that the estimated conditional expectation will likely not converge under the assumptions given. The same would happen for the Hermite family, and the problem is that the orthogonality properties of these families are difficult to exploit when it comes to implementing the LSM method.

An alternative choice of polynomial family are the shifted Chebyshev polynomials of the first or second kind which are orthogonal on the right interval. For example, the first kind of Chebyshev polynomials are orthogonal on the interval $(0;1)$ with respect to the weighting function $w(x) = (x - x^2)^{-1/2}$. However, though orthogonal on the right interval the weighting function results in unbounded regressors violating the assumptions underlying Lemma 1 in [22]. The last row in Table 8 shows the coefficients for the shifted Chebyshev polynomials of the second kind. This family of polynomials is again orthogonal on the interval of interest and now the weighting function is bounded. However, for this family of polynomials $\bar{U}_k(1) = k + 1$ and even though the polynomials are weighted by $w(x) = (x - x^2)^{1/2}$ it will not be the case that $\sup_{x \in \mathscr{X}} \|\bar{U}_K(x)\| \leq C \times K$, which again violates the assumptions underlying Lemma 1 in [22]. Thus, the highest rate with which the number of regressors is allowed to increased is likely lower for the Chebyshev polynomials than for the Legendre family.

Panel A of Table 9 provides results on the convergence of the estimated conditional expectations for option number 5 with the various polynomial families. The first thing to note from the table is that all the estimated coefficients are positive for the Laguerre and Hermite polynomials. Thus, the results indicate that when either of these families are used convergence is not obtained under the above

Table 8 Coefficients used to construct orthogonal polynomials recursively

Family	f_k	(a,b)	$w(x)$	a_{1k}	a_{2k}	a_{3k}	a_{4k}	$f_0(x)$	$f_1(x)$
Legendre	$\bar{P}_k(x)$	$(0,1)$	1	$k+1$	$-2k-1$	$4k+2$	k	1	$2x-1$
Laguerre	$L_k(x)$	$(0,\infty)$	$\exp(-x)$	$k+1$	$2k+1$	-1	k	1	$-x+1$
Hermite	$H_n(x)$	$(-\infty,\infty)$	$\exp(-x^2)$	1	0	2	$2k$	1	$2x$
Chebyshev 1	$\bar{T}_k(x)$	$(0,1)$	$(x-x^2)^{-1/2}$	1	-2	4	1	1	$2x-1$
Chebyshev 2	$\bar{U}_k(x)$	$(0,1)$	$(x-x^2)^{1/2}$	1	-2	4	1	1	$4x-2$

Notes: This table shows the coefficient used to construct the orthogonal polynomials used. The coefficients are the ones used in (36). A bar indicates that the polynomials are shifted. The interval over which the polynomials are orthogonal is given by (a,b) and $w(x)$ is the weighting function with respect to which they are orthogonal.

Table 9 Convergence results when using different polynomial families

Panel A: Conditional expectation estimates

Family	β_{BIAS^2}			β_{VAR}			β_{MSE}		
	Estim	S.E.	R^2	Estim	S.E.	R^2	Estim	S.E.	R^2
Legendre	−0.981	(0.004)	1.000	−0.704	(0.005)	1.000	−0.754	(0.002)	1.000
Laguerre	1.583	(0.819)	0.383	1.026	(0.551)	0.367	1.326	(0.667)	0.397
Hermite	0.155	(0.828)	0.006	0.066	(0.561)	0.002	0.166	(0.673)	0.010
Chebyshev 1	−0.981	(0.004)	1.000	−0.823	(0.049)	0.979	−0.847	(0.040)	0.987
Chebyshev 2	−0.981	(0.004)	1.000	−0.700	(0.003)	1.000	−0.753	(0.002)	1.000

Panel B: American option price estimates

Family	β_{BIAS^2}			β_{VAR}			β_{MSE}		
	Estim	S.E.	R^2	Estim	S.E.	R^2	Estim	S.E.	R^2
Legendre	−2.415	(0.820)	0.591	−0.998	(0.006)	1.000	−1.000	(0.006)	1.000
Laguerre	1.551	(1.271)	0.199	−0.856	(0.259)	0.645	0.313	(0.417)	0.086
Hermite	−0.623	(1.481)	0.029	−1.015	(0.011)	0.999	−0.400	(0.441)	0.120
Chebyshev 1	−1.334	(0.424)	0.623	−1.000	(0.007)	1.000	−1.000	(0.007)	1.000
Chebyshev 2	−2.011	(0.399)	0.809	−0.998	(0.007)	1.000	−0.999	(0.007)	1.000

Notes: This table reports the convergence results when using different polynomial families. Panel A reports the results from the regressions in (25)–(27) for the conditional expectation estimates. Panel B reports the results from the regressions in (29)–(31) for the American option price estimates. Results are presented for option number 5 (see the notes to Table 1 for the characteristics of the option).

assumptions.[9] For the Chebyshev polynomials of both kinds, on the other hand, the coefficients are estimated negatively which suggests that when these families are used convergence to the true conditional expectation is obtained. In fact, the results would seem to indicate that the convergence rate is significantly higher than the theoretical rates when using the Chebyshev polynomials of the first kind. However, care should be taken when interpreting the results as the assumptions are violated for this particular choice of regressors. Note that this may be the reason for the slightly lower explanatory power of the regressions with a $R^2 = 0.987$ indicating that the relationship between $\ln(N)$ and $\ln(MSE)$ is potentially nonlinear.

Panel B of Table 9 provides results for the convergence of the estimated American option prices for option number 5 with the various polynomial families. The table shows that the absence of convergence for the Laguerre polynomials results in price estimates which also do not converge to the true price. To be specific, the results show that whereas the variance does tend to zero as the number of paths is increased, this is not the case for the bias. It is due to this lack of convergence that the price estimate is non-convergent. For the Hermite polynomials the results would seem to suggest that the price estimate does converge though very slowly. However, again care has to be taken when interpreting these results as the very low R^2 indicates that the relationship between $\ln(N)$ and $\ln(MSE)$ is potentially nonlinear. Finally, for the Chebyshev polynomials, for which the estimated conditional expectation converges, the results indicate that the price estimates also converge and this with an estimated order insignificantly different from -1.

6.2 Other Types of Cross-sectional Regressions: The Unbounded Case

In [17] it is argued that in the cross-sectional regressions one should use only the in-the-money paths. The reason provided is that this allows a good fit with relatively few regressors, in particular, with fewer regressors than if all the paths are used. However, theoretically a much more important benefit is that it allows us to derive convergence rates using existing theory on semi-nonparametric series estimators for the cases analyzed above. The question remains, however, as to what can be obtained when all the paths are used in the cross-sectional regression. In particular, though the theoretical results are valid for the LSM method, the value function iteration method outlined in (8) and used by [7, 26] requires approximations not only for the in-the-money paths but for all the paths. In this subsection, we present results for both the extended LSM method which uses all the paths in the regressions and for the value function iteration method outlined in (8). We compare the performance

---

[9]In fact, the cross-sectional regressions break down in more than 80% of the cases when using eight Laguerre polynomials and in all the cases when using more than eight polynomials. The regressions also break down in all cases when using ten Hermite polynomials.

of the two methods as the number of early exercise times is increased since this is when differences can be expected to occur as shown in, e.g., [23]. In both cases, we use ordinary polynomials as regressors and the results can therefore be directly compared with the results for the regular LSM method in Table 7.

Panel A of Table 10 provides results on the convergence of the estimated conditional expectations for option number 5 with increasing number of exercise times when all the paths are used in the LSM method. The first thing to note from the table is that, though all the estimates are negative, the estimated order of convergence for the option with 25 exercise times is much lower than the corresponding estimate for the regular LSM method which uses only in-the-money paths in Table 7. The table shows that the reason for this lower estimate is the low estimates in absolute value for the variance term. For the value function iteration method, similar results are obtained for the variance term as Panel A of Table 11 shows. However, for this method the estimated values for the bias term is also much lower for the option with 25 exercise times. Finally, note that though the estimated order of convergence is numerically larger than -0.75 it does significantly decrease as the number of early exercises is increased.

Panel B of Table 10 provides results for the convergence of the estimated American option prices for option number 5 with increasing number of exercise times when all the paths are used in the LSM method. The table shows that, though the convergence rate for the conditional expectation estimates decreases, the estimated price still converges fast with a rate of N^{-1}. However, this is no longer the case for the value function iteration method as Panel B of Table 11 shows. In particular, for the option with 25 exercise times the estimated order of convergence is down to -0.77. The table also shows that this is caused by a slow convergence of the bias of the price estimate.

7 Conclusion

Recently, simulation methods have gained importance when it comes to American option pricing, and whereas it used to be considered difficult to price such products this is now known no longer to be the case. In particular, methods which combine the use of Monte Carlo simulation and regression techniques have been developed and used extensively. In this paper, we consider such methods and we examine numerically their convergence properties.

First of all, we provide numerical results complementing the recent theoretical work in [22] for the LSM method of [17]. The results confirm the theoretical rates derived for the simple two-period case with one underlying stochastic factor. They also show that in this situation the convergence rate can be increased towards a rate of N^{-1} by optimally picking the rate at which the number of paths is increased as a function of the number of regressors. We also examine the convergence rate of the estimated price, which are not available theoretically. The numerical results show

Table 10 Convergence results when using all the paths in the LSM method

Panel A: Conditional expectation estimates

Steps	β_{BIAS^2}			β_{VAR}			β_{MSE}		
	Estim	S.E.	R^2	Estim	S.E.	R^2	Estim	S.E.	R^2
2	−0.972	(0.011)	0.999	−0.950	(0.042)	0.988	−0.956	(0.031)	0.994
5	−0.974	(0.007)	1.000	−0.681	(0.010)	0.999	−0.790	(0.011)	0.999
10	−0.993	(0.009)	1.000	−0.658	(0.007)	0.999	−0.794	(0.012)	0.999
25	−1.005	(0.005)	1.000	−0.587	(0.005)	1.000	−0.742	(0.009)	0.999

Panel B: American option price estimates

Steps	β_{BIAS^2}			β_{VAR}			β_{MSE}		
	Estim	S.E.	R^2	Estim	S.E.	R^2	Estim	S.E.	R^2
2	−1.087	(0.485)	0.456	−1.003	(0.006)	1.000	−1.003	(0.006)	1.000
5	−1.574	(0.313)	0.808	−1.007	(0.008)	1.000	−1.008	(0.008)	1.000
10	−2.063	(0.242)	0.924	−1.020	(0.008)	1.000	−1.023	(0.008)	1.000
25	−1.676	(0.222)	0.905	−1.022	(0.004)	1.000	−1.025	(0.005)	1.000

Notes: This table reports the convergence results when using all the paths in the LSM method. Panel A reports the results from the regressions in (25)–(27) for the conditional expectation estimates. Panel B reports the results from the regressions in (29)–(31) for the American option price estimates. Results are presented for option number 5 (see the notes to Table 1 for the characteristics of the option).

Table 11 Convergence results when using value function iteration

Panel A: Conditional expectation estimates

Steps	β_{BIAS^2}			β_{VAR}			β_{MSE}		
	Estim	S.E.	R^2	Estim	S.E.	R^2	Estim	S.E.	R^2
2	−0.972	(0.011)	0.999	−0.950	(0.042)	0.988	−0.956	(0.031)	0.994
5	−1.175	(0.047)	0.991	−0.625	(0.024)	0.991	−0.923	(0.029)	0.994
10	−1.033	(0.067)	0.975	−0.619	(0.008)	0.999	−0.927	(0.046)	0.985
25	−0.854	(0.080)	0.950	−0.534	(0.026)	0.986	−0.820	(0.069)	0.959

Panel B: American option price estimates

Steps	β_{BIAS^2}			β_{VAR}			β_{MSE}		
	Estim	S.E.	R^2	Estim	S.E.	R^2	Estim	S.E.	R^2
2	−2.758	(0.755)	0.690	−0.998	(0.007)	1.000	−1.021	(0.010)	0.999
5	−1.225	(0.127)	0.939	−1.013	(0.009)	0.999	−1.087	(0.052)	0.986
10	−0.923	(0.113)	0.918	−1.016	(0.009)	1.000	−0.957	(0.081)	0.959
25	−0.713	(0.102)	0.891	−1.022	(0.008)	1.000	−0.769	(0.085)	0.931

Notes: This table reports the convergence results when using value function iteration. Panel A reports the results from the regressions in (25)–(27) for the conditional expectation estimates. Panel B reports the results from the regressions in (29)–(31) for the American option price estimates. Results are presented for option number 5 (see the notes to Table 1 for the characteristics of the option).

that this rate is insignificantly different from N^{-1}, which is the optimal rate one could hope for and the convergence rate of the corresponding simulated European price estimate.

Next, we generalize the numerical results to the situation with several underlying stochastic factors and to the situation with multiple early exercises. For the case with multiple stochastic factors, the results show that for smooth payoff functions convergence rates of the same order are found in the two-dimensional case. However, for options with non-smooth payoffs, such as options on the minimum, convergence rates are estimated to be lower. In terms of the price estimates the methods do converge fast, however, and the estimated order of convergence is close to -1 even for the minimum option. For the case with multiple early exercises, the results show that even with 25 exercise times convergence of the estimated conditional expectation is obtained and the rate is very close to the rate with only two exercise steps. For these options, the numerical results also show that the price converges fast.

Finally, we consider two situations which are not covered by the existing theory. We consider the case when the polynomial family used in the cross-sectional regressions is changed. The results show that care has to be taken in terms of the choice of regressors as the LSM method does not appear to converge when Laguerre or Hermite polynomials are used. Secondly, we consider the case when all the paths are used in the cross-sectional regressions which is necessary in applications of the value function iteration method proposed by [7, 26]. The results show that the rate at which the price converges deteriorates as the number of early exercises is increased when the value function iteration method is used. For example, when 25 exercise times are considered this rate is significantly lower than the rate of N^{-1} obtained with the LSM method.

This paper provides results on the convergence rates for simulation and regression methods for American option pricing. However, there are several interesting extensions for future research. First of all, from a theoretical point of view it remains of interest to obtain the actual convergence rates for the estimated conditional expectations in the multi-period situation. Secondly, considering the evidence provided in this paper on the fast convergence of the actual price estimates, it would be interesting to analyze under what setting it is possible to obtain these rates theoretically. Finally, since a general distributional theory is still not available for the simulation and regression methods for American option pricing, deriving this would also be an important area of future research.

8   Computational Issues

The actual implementation of the LSM method in this paper involves two important computational issues. The first is the generation of random normal variates and the second is related to the cross-sectional regressions performed. In this section, we explain how these two issues were dealt with in the present paper.

8.1 Random Number Generation

Generating "good" random numbers can be a quite difficult job. After all, random numbers generated by a computer are not random at all. Instead, conditional on the seed or the input to the random number generators (RNG) the sequence of random numbers is completely deterministic. Thus, it becomes of importance to control the seeds judiciously. The reason for this is that as we increase the number of paths from say 10,000 to 20,000 we want the first 10,000 paths of the batch of 20,000 paths to be the same. Otherwise, any arguments along the lines of what will happen as N tends to infinity are meaningless.

Our approach sets the seeds prior to any simulations and then simply re-uses these seeds in all calls to the random number generator for that particular iteration. Unfortunately, for some RNG it is not possible to initialize these seeds in such a way that it is ensured that the sequences of random variates are nonoverlapping. This could easily be done with, say, a simple linear congruential generator (see [13]). Instead, we have to make a number of actual calls to the RNG. For example, in the first application where we have two time periods we need to generate 2,000 sets of seeds since we have a total of 1,000 iterations. As we need at most 100,000 random variates, we generate this number of random numbers 2,000 times and store the seeds used by the RNG at each step. These 2,000 seeds are then used in the simulation of the stock prices ensuring that, e.g., the first stock path is the same irrespective of the total number of paths.

A somewhat related issue is how to add early exercise times and, in particular, how to generate the relevant random numbers when adding, e.g., the second early exercise point. In this paper these numbers are generated sequentially starting at the time of maturity, an approach which fits well since we generate the paths using a Brownian bridge. Thus, the first batch of random numbers are always used to generate the stock paths at maturity and thus the European option prices are the same irrespective of how many early exercise times we are considering. The next batch is then used to generate the paths at time $T - 1$, i.e., the last early exercise point. In this way we continue until random numbers have been generated for all early exercise times.

8.2 Cross-sectional Regressions

As pointed out the most important element of the convergence theorem is to establish conditions under which the number of regressors, M, can tend to infinity along with the number of simulated paths, N. In particular, the requirement that $M^3/N \to 0$ is what ensures that the second moment matrix is nonsingular. However, while this may be true in theory, numerical singularities may occur in an actual implementation. In principle, we would like to use the most stable of the routines provided by the particular software. However, it may nevertheless be important to control these procedures. For example, if this software provided function breaks down it has to be determined if this is caused by actual second moment matrix singularities or if it is in fact caused by what only appears to be a numerically singularity.

The procedure used in this paper is to compute the singular value decomposition (SVD) of the matrix of regressors and compute the ratio between the smallest and the largest singular values. Following [20], if this ratio is less than 10^{-12} the matrix is said to be ill conditioned and we refrain from performing the cross-sectional regressions. Note that this is done whenever the software-specific function fails only, and in these situations we report a β of zero's. We count the number of situations and report the fraction of regressions where the regression procedure broke down.

References

1. Abramowitz, M., Stegun, IA.: Handbook of Mathematical Functions. Dover Publications, New York (1970)
2. Barraquand, J., Martineau, D.: Numerical valuation of high dimensional multivariate American securities. J. Financ. Quant. Anal. **30**, 383–405 (1995)
3. Boyle, P.P.: A lattice framework for option pricing with two state variables. J. Financ. Quant. Anal. **23**, 1–12 (1988)
4. Boyle, P.P., Evnine, J., Gibbs, S.: Numerical evaluation of multivariate contingent claims. Rev. Financ. Stud. **2**, 241–250 (1989)
5. Broadie, M., Glasserman, P.: Pricing American-style securities using simulation. J. Econ. Dynam. Contr. **21**, 1323–1352 (1997)
6. Broadie, M., Glasserman, P.: A stochastic mesh method for pricing high-dimensional American options. J. Comput. Finance. **7**(4), 35–72 (2004)
7. Carriere, J.F.: Valuation of the early-exercise price for options using simulations and nonparametric regression. Insur. Math. Econ. **19**, 19–30 (1996)
8. Chow, Y.S., Robbins, H., Siegmund, D.: Great Expectations: The Theory of Optimal Stopping. Houghton Mifflin, New York (1971)
9. Clément, E., Lamberton, D., Protter, P.: An analysis of the Longstaff-Schwartz algorithm for American option pricing. Finance Stochast. **6**(4), 449–471 (2002)
10. Duffie, D.: Dynamic Asset Pricing Theory. Princeton University Press, Princeton, New Jersey (1996)
11. Feinerman, R.P., Newman, D.J.: Polynomial Approximation. Williams and Wilkins, Baltimore, MD (1973)
12. Gerhold, S.: The Longstaff-Schwartz algorithm for Lévy models: results on fast and slow convergence. Ann. Appl. Probab. 21(2):589–608 (2011)
13. Glasserman, P.: Monte Carlo Methods in Financial Engineering. Springer-Verlag, New York, Inc., New York, USA (2004)
14. Glasserman, P., Yu, B.: Number of paths versus number of basis functions in American option pricing. Ann. Appl. Probab. **14**(4), 2090–2119 (2004)
15. Judd, K.L.: Numerical Methods in Economics. MIT, Cambridge, Massachusetts (1998)
16. Karatzas, I.: On the pricing of American options. Appl. Math. Optim. **17**, 37–60 (1988)
17. Longstaff, F.A., Schwartz, E.S.: Valuing American options by simulation: a simple least-squares approach. Rev. Financ. Stud. **14**, 113–147 (2001)
18. Moreno, M., Navas, J.F.: On the robustness of least-squares Monte Carlo (lsm) for pricing American options. Rev. Derivatives Res. **6**, 107–128 (2003)

19. Pagan, A., Ullah, A.: Nonparametric Econometrics. Cambridge University Press, Cambridge, UK (1999)
20. Press, W.H., Teukolsky, S.A., Vetterling, W.T., Flannery, B.P.: Numerical Recipies in C: The Art of Scientific Computing. Cambridge University Press, Cambridge (1997)
21. Stentoft, L.: Assessing the least squares Monte-Carlo approach to American option valuation. Rev. Derivatives Res. 7(3), 129–168 (2004a)
22. Stentoft, L.: Convergence of the least squares Monte Carlo approach to American option valuation. Manag. Sci. 50(9), 1193–1203 (2004b)
23. Stentoft, L.: Value function approximation or stopping time approximation: a comparison of two recent numerical methods for American option pricing using simulation and regression. Forthcoming in Journal of Computational Finance
24. Stentoft, L.: "American option pricing using simulation: An introduction with an application to the GARCH option pricing model", Handbook of research methods and applications in empirical finance, Adrian Bell, Chris Brooks, Marcel Prokopczuk, eds., Edward Elgar Publishing, 2012
25. Tilley, J.A.: Valuing American options in a path simulation model. Trans. Soc. Actuaries, Schaumburg XLV:499–520 (1993)
26. Tsitsiklis, J.N., Van Roy, B.: Regression methods for pricing complex American-style options. IEEE Trans. Neural Network. 12(4), 694–703 (2001)

The COS Method for Pricing Options Under Uncertain Volatility

M.J. Ruijter and C.W. Oosterlee

Abstract We develop a method for pricing financial options under uncertain volatility. The method is based on the dynamic programming principle and a Fourier cosine expansion method. Local errors in the vicinity of domain boundaries, originating from the use of Fourier series expansions, may hamper the algorithm's convergence. We use an extrapolation method to deal with these errors.

1 Introduction

A Fourier method for solving stochastic control problems has been developed in [12]. This paper is a conference proceedings version, in which we discuss a practical application in detail, i.e., the pricing of an option under uncertain volatility.

In [11], an implicit discretization method for the governing partial differential equation (PDE) to solve a similar problem has been applied. Numerical experiments were performed using the butterfly spread and digital options. It was demonstrated that a non-monotone scheme may lead to incorrect, nonviscosity solutions. The authors in [13] combined an exponentially fitted finite volume method for the space direction and an implicit scheme in time for the PDE. The method was tested by using butterfly spread, double barrier call, and digital call options. The option prices in these papers will serve as reference to which we will compare our results.

M.J. Ruijter (✉)
Centrum Wiskunde & Informatica, Science Park 123, Amsterdam, The Netherlands

CPB Netherlands Bureau for Economic Policy Analysis, Van Stolkweg 14,
Den Haag, The Netherlands
e-mail: m.j.ruijter@cwi.nl

C.W. Oosterlee
Centrum Wiskunde & Informatica, Science Park 123, Amsterdam, The Netherlands

Delft University of Technology, Mekelweg 4, Delft, The Netherlands
e-mail: c.w.oosterlee@cwi.nl

M. Cummins et al. (eds.), *Topics in Numerical Methods for Finance*, Springer Proceedings
in Mathematics & Statistics 19, DOI 10.1007/978-1-4614-3433-7_6,
© Springer Science+Business Media New York 2012

Our method relies on the dynamic programming principle and the COS formula [4], which is based on Fourier cosine series expansions. A recursive algorithm is developed based on the recursive recovery of the series coefficients, see Sect. 2.

It is known that Fourier cosine expansions may be inaccurate near spatial boundaries, particularly outside the expansion interval. These errors may propagate backwards in time. We give insights into the source of these local errors in Sect. 3.1. Based on this, we propose an extrapolation technique near the domain boundaries as an accurate solution technique in this context (Sect. 3.2).

In Sect. 4, we test the pricing method on digital and bull split-strike combo options. Besides, we analyze the performance of the COS method under discontinuous density functions. Finally, Sect. 5 summarizes and concludes.

2 Problem Description and Pricing Method

Let $T > 0$ be a finite terminal time, $((\Omega, \mathscr{F}, \mathscr{F}_s)_{0 \leq s \leq T}, P)$ a complete probability space satisfying the usual conditions, W_s an $\mathscr{F}_s$-adapted one-dimensional Brownian motion, and q_s an $\mathscr{F}_s$-adapted Poisson process with intensity λ.

The model we use is described in [11, 13]. The risk-neutral dynamics of the asset price is assumed to evolve according to either a geometric Brownian motion (GBM),

$$\mathrm{d}S_s = rS_s\mathrm{d}s + \alpha_s S_s \mathrm{d}W_s, \tag{1}$$

or Merton's jump-diffusion process,

$$\mathrm{d}S_s = (r - \lambda\kappa)S_s\mathrm{d}s + \alpha_s S_s\mathrm{d}W_s + (\mathrm{e}^J - 1)S_s\mathrm{d}q_s. \tag{2}$$

Here, $\kappa := \mathbb{E}[\mathrm{e}^J - 1]$ and r is the risk-neutral interest rate. The jumps J are normally distributed with mean μ_J and standard deviation σ_J. $(\alpha_s)_{0 \leq s \leq T}$ represents an uncertain volatility process, which is valued in the interval $[\alpha^-, \alpha^+]$. We consider the worst case for an investor with a long position in a European-style option. Then, the option value at time t reads:

$$v(t, S) = \inf_{\alpha \in \mathscr{A}} \mathbb{E}^{t,S}[\mathrm{e}^{-r(T-t)}g(S_T)], \tag{3}$$

where $g(.)$ is a prescribed payoff function. $\mathscr{A}$ denotes the set of admissible control processes [10]. The pricing problem is now formulated as a stochastic control problem, whereby the process α_s is the control process, with values in the bounded set $A = [\alpha^-, \alpha^+]$.

The corresponding Hamilton–Jacobi–Bellman (HJB) equations read

$$-\frac{\partial v}{\partial t}(t, S) + rv(t, S) = \min_{\alpha \in [\alpha^+, \alpha^-]} \left[rS\frac{\partial v}{\partial S}(t, S) + \frac{1}{2}\alpha^2 S^2 \frac{\partial^2 v}{\partial S^2}(t, S) \right],$$

$$\forall (t, S) \in [0, T) \times \mathbb{R}_+, \tag{4}$$

under GBM, and

$$
-\frac{\partial v}{\partial t}(t,S) + rv(t,S)
$$

$$
= \min_{\alpha \in [\alpha^+, \alpha^-]} \left[(r - \lambda \kappa) S \frac{\partial v}{\partial S}(t,S) + \frac{1}{2} \alpha^2 S^2 \frac{\partial^2 v}{\partial S^2}(t,S) + \lambda e[v(t, e^J S) - v(t,S)] \right],
$$

$$
\forall (t,S) \in [0,T) \times \mathbb{R}_+, \tag{5}
$$

under the jump-diffusion process. This yields

$$
\text{if } \frac{\partial^2 v}{\partial S^2} \leq 0 \Rightarrow \text{take } \alpha = \alpha^+,
$$

$$
\text{if } \frac{\partial^2 v}{\partial S^2} > 0 \Rightarrow \text{take } \alpha = \alpha^-, \tag{6}
$$

which allows us to restrict the set of possible control values to $A = \{\alpha^-, \alpha^+\}$. We see that the control value, that is the volatility, is a function of the Greek $\Gamma = \partial^2 v / \partial S^2$.

Remark 1. Stochastic control problems may be solved employing numerical PDE techniques to the corresponding HJB equation. We refer to [6, 11] for numerical discretization methods. Then, issues about convergence to the correct, viscosity solution arise. The viscosity solution concept was introduced by P.L. Lions [9]. We refer to [2] for a general introduction to viscosity solutions and some general uniqueness and existence results. As we use the dynamic programming approach, we will not go into details about this.

We switch to log-asset price processes, $X_s := \log S_s$, that belong to the class of Lévy processes. For GBM, we then deal with the Brownian motion

$$
dX_s = \left(r - \frac{1}{2} \alpha_s^2 \right) ds + \alpha_s dW_s, \tag{7}
$$

whereas the log-jump-diffusion process reads

$$
dX_s = \left(r - \lambda \kappa - \frac{1}{2} \alpha_s^2 \right) ds + \alpha_s dW_s + J dq_s. \tag{8}
$$

2.1 The COS Method

In this section, we set up a general method to solve the pricing problem under a one-dimensional *Lévy process*, X_t. The method is based on the dynamic programming principle and uses the so-called COS formula, which was developed in [4] for pricing European options. It results in a recursive algorithm based on the Fast Fourier Transform algorithm. We will explain the COS formula in Sect. 2.2. We start here with the discrete-time framework of the solution method.

We deal with a discrete-time stochastic control problem, with $\mathcal{M}$ control times. Convergence of the numerical solution to the solution of the original problem (3) is achieved by increasing the number of time steps. Initial time is denoted by t_0. We take a fixed equidistant grid of control times $t_0 < t_1 < \cdots t_m < \cdots < t_{\mathcal{M}} = T$, with $\Delta t := t_{m+1} - t_m$. As a discrete approximation, we assume that the volatility process is constant during the time intervals $[t_m, t_{m+1}]$. At each control time t_m, with $m < \mathcal{M}$, one can choose a control value from the set $A = \{\alpha^-, \alpha^+\}$, which influences the stochastic process during the time interval $[t_m, t_{m+1}]$. This value is denoted by $\alpha_m \in \{\alpha^-, \alpha^+\}$, where the subscript refers to the control time. The choice depends on the current asset price. With this notation, bold-faced $\boldsymbol{\alpha}$ denotes a control process and α_m denotes a single control value.

The value function reads

$$v(t,x) := \min_{\boldsymbol{\alpha} \in \mathscr{A}} \mathbb{E}^{t,x} \left[e^{-r(T-t)} g(e^{X_T}) \right]. \tag{9}$$

$\hat{\mathscr{A}} \subset \mathscr{A}$ denotes the set of all possible control paths $\{\alpha_m\}_{m=0}^{\mathcal{M}-1}$, where α_m is valued in the control set A.

The dynamic programming principle gives:

$$v(t_{m-1}, x) = \min_{\alpha_{m-1} \in A} e^{-r\Delta t} \mathbb{E}[v(t_m, X_{t_m}) | X_{t_{m-1}} = x, \alpha_{m-1}]$$

$$= \min \left[c(t_{m-1}, x, \alpha^-), c(t_{m-1}, x, \alpha^+) \right]. \tag{10}$$

The expectation $c(t_{m-1}, x, \alpha)$ is called the *continuation value* under control α.

2.2 COS Formula

In this section, we explain briefly our method to approximate the continuation value. The numerical method is based on series expansions of the value function at a next time level and the density function. For more details, we refer to [12].

We start with truncation of the infinite integration range of the expectation to some interval $[a,b] \subset \mathbb{R}$,

$$c(t_{m-1}, x, \alpha) \approx e^{-r\Delta t} \int_a^b v(t_m, y) f(y|x, \alpha) dy. \tag{11}$$

Next, we use the Fourier cosine series expansions of the density function and the value function on $[a,b]$, with series coefficients

$$A_k(x, \alpha) := \frac{2}{b-a} \int_a^b f(y|x, \alpha) \cos \left(k\pi \frac{y-a}{b-a} \right) dy$$

$$\text{and } V_k(t_m) := \frac{2}{b-a} \int_a^b v(t_m, y) \cos \left(k\pi \frac{y-a}{b-a} \right) dy, \tag{12}$$

respectively. The density function of a stochastic process is usually not known, but often its characteristic function is known (see [3, 4]). The coefficients $A_k(x, \alpha)$ can be approximated as follows

$$A_k(x, \alpha) \approx \frac{2}{b-a} \text{Re} \left(\varphi \left(\frac{k\pi}{b-a} \middle| x, \alpha \right) e^{-ik\pi \frac{a}{b-a}} \right) := F_k(x, \alpha). \tag{13}$$

Re (.) denotes taking the real part of the input argument. $\varphi(.|x, \alpha)$ is the conditional *characteristic function* of X_{t_m}, given $X_{t_{m-1}} = x$ and $\alpha_{m-1} = \alpha$. For Lévy processes, it follows that:

$$\varphi(u|x, \alpha) = \varphi(u|0, \alpha)e^{iux} := \varphi_{\text{levy}}(u|\alpha)e^{iux}. \tag{14}$$

For Brownian motion, Eq. (7), the characteristic function reads

$$\varphi_{\text{levy}}(u|\alpha) = \exp \left(iu \left(r - \frac{1}{2}\alpha^2 \right) \Delta t - \frac{1}{2}u^2\alpha^2 \Delta t \right) \tag{15}$$

and for the log-jump-diffusion process, Eq. (8), we have

$$\varphi_{\text{levy}}(u|\alpha) = \exp \left(iu \left(r - \lambda\kappa - \frac{1}{2}\alpha^2 \right) \Delta t - \frac{1}{2}u^2\alpha^2 \Delta t \right) e^{\lambda\Delta t (\exp(i\mu_J u - \frac{1}{2}u^2\sigma_J^2)-1)}. \tag{16}$$

Inserting the Fourier cosine coefficients, combined with a truncation of the series summations, gives us the *COS formula* for approximation of $c(t_{m-1}, x, \alpha)$:

$$\hat{c}(t_{m-1}, x, \alpha|[a,b], N) := e^{-r\Delta t} \sum_{k=0}^{N-1} {}' \text{Re} \left(\varphi_{\text{levy}} \left(\frac{k\pi}{b-a} \middle| \alpha \right) e^{ik\pi \frac{x-a}{b-a}} \right) V_k(t_m). \tag{17}$$

$\sum'$ in (17) indicates that the first term in the summation is weighted by one-half.

2.3 Recursion Formula for Coefficients $V_k(t_m)$

The algorithm for solving the stochastic control problem (9) is based on the recursive recovery of the coefficients $V_k(t_m)$, starting with the coefficients at the terminal time, $V_k(T)$, for which often an analytic solution is available. These coefficients are used for the approximation of the continuation value at time $t_{\mathcal{M}-1}$.

Next we consider the coefficients that are used to approximate the continuation value at time t_{m-1}, for $m \leq \mathcal{M} - 1$. The value function at time t_m appears in the terms $V_k(t_m)$ and we need to find an optimal volatility for all state values $y \in [a, b]$. We determine sub-domains $\mathcal{D}_m^+$ and $\mathcal{D}_m^- \subset [a, b]$, so that for each $y \in \mathcal{D}_m^\pm$ it is optimal to choose control $\alpha_m^\pm$ at control time t_m. The subscript of $\alpha_m^\pm$ indicates the time level and the superscript represents the control value. We split the integral for the definition of V_k into different parts:

$$V_k(t_m) = \frac{2}{b-a} \int_{\mathscr{D}_m^+} c(t_m, y, \alpha_m^+) \cos\left(k\pi \frac{y-a}{b-a}\right) dy$$

$$+ \frac{2}{b-a} \int_{\mathscr{D}_m^-} c(t_m, y, \alpha_m^-) \cos\left(k\pi \frac{y-a}{b-a}\right) dy$$

$$:= C_k(t_m, \mathscr{D}_m^+, \alpha_m^+) + C_k(t_m, \mathscr{D}_m^-, \alpha_m^-), \quad (m \neq \mathscr{M}). \tag{18}$$

We approximate the terms $\hat{C}_k$ at time t_{M-1} by using the COS formula for $c(t_{\mathscr{M}-1}, y, \alpha)$, i.e., Eq. (17). Interchanging summation and integration then gives the approximation

$$\hat{C}_k(t_{\mathscr{M}-1}, z_1, z_2, \alpha) = e^{-r\Delta t} \text{Re}\left(\sum_{j=0}^{N-1}{}' \varphi_{\text{levy}}\left(\frac{j\pi}{b-a}\Big|\alpha\right) V_j(t_{\mathscr{M}}) M_{k,j}(z_1, z_2)\right), \tag{19}$$

where the elements of matrix $M(z_1, z_2)$ are given by:

$$M_{k,j}(z_1, z_2) := \frac{2}{b-a} \int_{z_1}^{z_2} e^{ij\pi \frac{y-a}{b-a}} \cos\left(k\pi \frac{y-a}{b-a}\right) dy. \tag{20}$$

The parameters of the matrices M are the boundary values of their respective integration ranges. For the coefficients $V_k(t_m)$, $1 \leq m \leq \mathscr{M} - 2$, the approximations $\hat{c}(t_m, y, \alpha)$ and $\hat{V}_k(t_{m+1})$ will be used to approximate the terms $C_k(t_m, z_1, z_2, \alpha)$, which yields

$$\hat{C}_k(t_m, z_1, z_2, \alpha) = e^{-r\Delta t} \text{Re}\left(\sum_{j=0}^{N-1}{}' \varphi_{\text{levy}}\left(\frac{j\pi}{b-a}\Big|\alpha\right) \hat{V}_j(t_{m+1}) M_{k,j}(z_1, z_2)\right). \tag{21}$$

We get the following numerical approximation of the Fourier cosine coefficients:

$$\hat{V}(t_m) = e^{-r\Delta t} \text{Re}\left(M(\mathscr{D}_m^+)\Lambda^+ \hat{V}(t_{m+1})\right) + e^{-r\Delta t} \text{Re}\left(M(\mathscr{D}_m^-)\Lambda^- \hat{V}(t_{m+1})\right),$$

$$m = 1, \ldots, \mathscr{M} - 2, \tag{22}$$

where $\Lambda^\pm$ is a diagonal matrix with entries

$$\hat{\mathbf{w}}^\pm = \left\{\hat{w}_j^\pm\right\}_{j=0}^{N-1} \quad \text{with} \quad \hat{w}_j^\pm = \varphi\left(\frac{j\pi}{b-a}\Big|\alpha_m^\pm\right), \quad \hat{w}_0^\pm = \frac{1}{2}\varphi(0|\alpha_m^\pm). \tag{23}$$

An additional error is introduced because the coefficients are approximated using the approximated elements $\hat{V}_j(t_{m+1})$. We propose an accurate alternative approximation for the Fourier coefficients $V_k(t_m)$ in Sect. 3.2.

The matrix-vector products $M\mathbf{w}$ in the terms $\hat{C}$ can be computed by a Fourier-based algorithm, as stated in the next result:

Result 1. [Efficient computation of $\hat{C}(t_m, z_1, z_2, \alpha)$] [5]. The matrix-vector product $M(z_1, z_2)\mathbf{w}$ can be computed in $O(N \log_2 N)$ operations, with the help of the Fast Fourier Transform (FFT) algorithm.

The pricing problem is solved backwards in time, starting with the coefficients $V_k(t_{\mathscr{M}})$. The terms $\hat{V}_k(t_m)$ are recovered recursively by determining the sub-domains $\mathscr{D}_m^{\pm}$ and using the FFT algorithm. Finally, we compute $\hat{v}(t_0, x_0)$ by inserting $\hat{V}_k(t_1)$ into Eq. (17). The computational complexity of the algorithm is $O(\mathscr{M} N \log_2 N)$, as we need to compute $\mathscr{M}$ time steps.

3   Error Analysis

In this section, we analyze the error of the COS method for pricing options under uncertain volatility and base our analysis on [4, 5, 12]. Errors in the COS method are introduced by the COS formula and by evolution through time via the coefficients $\hat{V}_k$ and a possibly incorrect control α. We start with the local error in Sect. 3.1. The local error may be significant in the vicinity of the boundaries. This may give difficulties during the recursive recovery of the Fourier cosine coefficients $V_k(t_m)$. In Sect. 3.2, we propose an improved approximation for $V_k(t_m)$, which is more accurate than $\hat{V}_k(t_m)$ from (22). Finally, the propagating error in the backward recursion is analyzed (Sect. 3.3).

3.1   Local Error COS Formula

We define the local error of the COS formula for the continuation value by

$$\varepsilon_{\mathrm{COS}}(t_{m-1}, x, \alpha | [a, b], N) := c(t_{m-1}, x, \alpha) - \hat{c}(t_{m-1}, x, \alpha | [a, b], N). \qquad (24)$$

We first assume that the terms $V_k(t_m)$ are exact. The error consists of three parts. The integration range truncation error enters by truncation of the infinite domain to the finite domain $[a, b]$. Conversely, the error related to approximating $A_k(x, \alpha)$ by $F_k(x, \alpha)$ (Eq. (13)) is due to replacing the finite domain by an infinite domain in Eq. (13). Addition of both errors gives

$$e^{-r\Delta t} \int_{\mathbb{R} \setminus [a, b]} [v(t_m, y) - \hat{v}(t_m, y)] f(y | x, \alpha) dy. \qquad (25)$$

The series truncation error on $[a, b]$ reads

$$\frac{b-a}{2} e^{-r\Delta t} \sum_{k=N}^{+\infty} A_k(x, \alpha) V_k(t_m) = e^{-r\Delta t} \int_a^b v(t_m, y) f(y|x, \alpha) - \hat{v}(t_m, y) \hat{f}(y|x, \alpha) dy.$$

(26)

The functions $\hat{v}(t_m, y)$ and $\hat{f}(y|x, \alpha)$ denote the Fourier cosine series expansions of the value function and the density function, using N terms in the series summations. The convergence rate of Fourier cosine series depends on the properties of the approximated functions in the expansion interval. Information about different convergence types can be found in [1]. Based on that theory, we find that the series truncation error converges exponentially for density functions in the class $C^\infty([a, b])$. A density function with discontinuity in one of its derivatives results in an algebraic convergence.

We can write

$$\hat{c}(t_{m-1}, x, \alpha | [a, b], N)$$

$$= e^{-r\Delta t} \int_a^b \hat{v}(t_m, y) \hat{f}(y|x, \alpha) dy + e^{-r\Delta t} \int_{\mathbb{R} \backslash [a, b]} \hat{v}(t_m, y) f(y|x, \alpha) dy. \quad (27)$$

Remark 2. Note that the Fourier cosine series expansions used in Sect. 2.2 are defined for $y \in [a, b]$, whereas the function $\hat{v}(t_m, y)$ is evaluated on $\mathbb{R} \backslash [a, b]$. Here we denote by function $\hat{v}(t_m, y)$, for $y \in \mathbb{R} \backslash [a, b]$, the symmetric extension of the Fourier cosine series expansion outside the expansion interval. This value will usually be different from $v(t_m, y)$, even if N tends to infinity.

If, for given x, the integration interval $[a, b]$ is chosen sufficiently wide, the series truncation error dominates the overall local error. This implies that for smooth density functions the local error converges exponentially to zero, otherwise it goes algebraically.

For a given interval $[a, b]$, the local error may, however, be large if x is in the vicinity of the domain boundaries. A local error may propagate via the backward recursion.

3.2 Improvement by Extrapolation

The coefficients $V_k(t_m)$ are recovered recursively, backwards in time. In that case, the local error, ε_{COS}, described in the previous section, may propagate through time. Here, we propose a technique to deal with the inaccurate approximation near domain boundaries. The idea is to determine the area in which inaccurate approximate values from the COS method occur. In this area, we employ an *extrapolation technique* to compute a value with improved accuracy, using accurate numerical continuation values from the neighboring region.

In practical applications, it may be possible to determine the area in which $\hat{c}(t_{m-1}, x, \alpha)$ is inaccurate, assuming that coefficients $V_k(t_m)$ are exact. The density function, together with the value function, gives the desired information. For instance, suppose we can calculate a value x^*, so that the continuation value is well approximated for $x \in [a, x^*]$ and is inaccurate for $x \in [x^*, b]$.[1] The continuation function $c(t_{m-1}, x, \alpha)$, on $[x^*, b]$, can then be approximated by an extrapolation technique. For this, we employ a second-order Taylor expansion in x^*:

$$c^{\text{ex}}(t_{m-1}, x, \alpha) := \hat{c}(t_{m-1}, x^*, \alpha) + \hat{c}_x(t_{m-1}, x^*, \alpha)(x - x^*) + \frac{1}{2}\hat{c}_{xx}(t_{m-1}, x^*, \alpha)(x - x^*)^2. \tag{28}$$

The derivatives can easily be computed in this setting, as:

$$\hat{c}_x(t_{m-1}, x^*, \alpha) = e^{-r\Delta t} \sum_{k=0}^{N-1}{}' \text{Re}\left(\varphi_{\text{levy}}\left(\frac{k\pi}{b-a}\Big|\alpha\right) e^{ik\pi\frac{x^*-a}{b-a}} \frac{ik\pi}{b-a}\right) V_k(t_m), \tag{29}$$

$$\hat{c}_{xx}(t_{m-1}, x^*, \alpha) = e^{-r\Delta t} \sum_{k=0}^{N-1}{}' \text{Re}\left(\varphi_{\text{levy}}\left(\frac{k\pi}{b-a}\Big|\alpha\right) e^{ik\pi\frac{x^*-a}{b-a}} \left(\frac{ik\pi}{b-a}\right)^2\right) V_k(t_m). \tag{30}$$

We denote the extrapolated continuation value by

$$\tilde{c}(t_{m-1}, x, \alpha) := \begin{cases} \hat{c}(t_{m-1}, x, \alpha), & \text{for } x \in [a, x^*], \\ c^{\text{ex}}(t_{m-1}, x, \alpha), & \text{for } x \in [x^*, b]. \end{cases} \tag{31}$$

The local error of the COS formula *with* extrapolation technique is denoted by

$$\tilde{\varepsilon}_{\text{COS}}(t_{m-1}, x, \alpha | [a, b], N) := c(t_{m-1}, x, \alpha) - \tilde{c}(t_{m-1}, x, \alpha | [a, b], N), \tag{32}$$

and we have:

$$\tilde{\varepsilon}_{\text{COS}}(t_{m-1}, x, \alpha | [a, b], N) = O((x - x^*)^3), \quad \text{for } x \in [a, x^*]. \tag{33}$$

We use continuation value $\tilde{c}$ to determine the optimal control law and to approximate the terms C_k by

$$\tilde{C}_k(t_{m-1}, z_1, z_2, \alpha) := \frac{2}{b-a} \int_{z_1}^{z_2} \tilde{c}(t_{m-1}, y, \alpha) \cos\left(k\pi\frac{y-a}{b-a}\right) dy. \tag{34}$$

The corresponding Fourier coefficients are denoted by $\tilde{V}_k(t_{m-1})$.

---

[1] The methodology above can also be applied if the approximated continuation value is inaccurate in a certain area $[a, x^{**}]$.

3.3 Error Propagation in the Backward Recursion

Next we discuss the error convergence if we employ the extrapolation methodology from Sect. 3.2. We define

$$\varepsilon_k(t_m) := V_k(t_m) - \tilde{V}_k(t_m).\qquad (35)$$

The error of the approximated Fourier coefficients now converges exponentially in N under certain conditions [12]:

Result 2. With a sufficiently accurate extrapolation technique, with $[a,b] \subset \mathbb{R}$ chosen sufficiently wide and a probability density function f in $C^\infty([a,b])$, error $\varepsilon_k(t_m)$ converges exponentially in N for $1 \le m \le \mathcal{M} - 1$.

The proof of this result is similar to that for pricing Bermudan options, which can be found in [5]. It can also be proved that if the local error converges algebraically, then so does $\varepsilon_k(t_m)$.

4 Numerical Experiments

We test the COS method by pricing digital and bull split-strike combo (bull) options. The payoff of a digital call (put) option is given by

$$g(S) = \begin{cases} 1 & (0), \text{ for } S > K, \\ 0 & (1), \text{ for } S < K, \end{cases}\qquad (36)$$

for certain strike price K. The bull option is a combination of a short position in a put option with strike K_1 and a long position in a call with strike K_2, where $K_1 < K_2$. The payoff function is given by

$$g(S) = \begin{cases} S - K_1, & \text{for } S \le K_1, \\ 0, & \text{for } K_1 \le S \le K_2, \\ S - K_2, & \text{for } K_2 \le S. \end{cases}\qquad (37)$$

For the tests, we take $K = 100$, $K_1 = 90$, and $K_2 = 110$. Figure 1 shows the corresponding payoff functions. We switch to the log-asset price domain. The terminal coefficients for the digital call option are given by

$$V_k(t_{\mathcal{M}}) = \frac{2}{b-a} \psi_k(\log K, b, a, b) \qquad (a \le \log K \le b)\qquad (38)$$

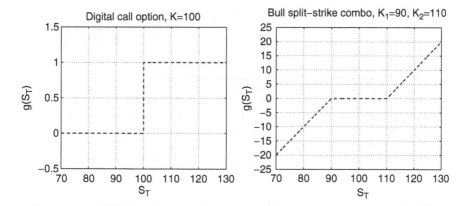

Fig. 1 Payoff digital call and bull option

and for the bull split-strike combo option we have

$$V_k(t_\mathcal{M}) = \frac{2}{b-a}\left[\chi_k(\log K_2, b, a, b) - \psi_k(\log K_2, b, a, b) - \psi_k(\log K_1, b, a, b)\right.$$
$$\left. + \chi_k(\log K_1, b, a, b)\right] \quad (a \leq \log K_1, \log K_2 \leq b). \tag{39}$$

For the functions χ_k and ψ_k, we refer to [4].

We use the following model parameters

$$T = 0.25, \quad r = 0.05, \quad S_0 = 100 \tag{40}$$

and for the jumps

$$\lambda = 0.01, \quad \mu_J = -0.90, \quad \text{and} \quad \sigma_J = 0.45. \tag{41}$$

4.1 Fixed Volatility Experiment

We start with a test case with a fixed, known volatility, $\sigma = 0.15$, so that the problem simplifies to pricing a European option.

As in [4], we take

$$[a, b] = \left[x_0 + \xi_1 - L\sqrt{\xi_2 + \sqrt{\xi_4}}, x_0 + \xi_1 + L\sqrt{\xi_2 + \sqrt{\xi_4}}\right], \quad L \in [8, 10]. \tag{42}$$

Table 1 Results option pricing method with fixed volatility

(a) Digital put option $v(t_0,x_0)=0.443$ (GBM, $L=10$)		(b) Bull option $v(t_0,x_0)=0.356$ (GBM, $L=10$)		(c) Digital put option $v(t_0,x_0)=0.443$ (Jump diffusion, $L=8$)		
N	Error	N	Error	N	$\hat{v}(t_0,x_0)$	Error
4	$-3.22e-02$	4	$1.43e+00$	4	0.4915156	$-1.04e-01$
8	$-1.36e-02$	8	$1.42e-01$	8	0.4825689	$-9.54e-02$
16	$-8.96e-04$	16	$1.01e-02$	16	0.4606004	$-7.34e-02$
32	$-4.79e-08$	32	$2.97e-08$	32	0.4253037	$-3.81e-02$
64	$1.67e-16$	64	$-1.08e-14$	64	0.3928054	$-5.65e-03$
128	$1.67e-16$	128	$-1.08e-14$	128	0.3871621	$-9.10e-06$

For the cumulants ξ_1, ξ_2, and ξ_4 of the Brownian motion and the Merton process, we refer to [4]. If N, the number of terms in the cosine expansion, is chosen sufficiently large, then a larger computational domain should not affect the option price.

4.1.1 Geometric Brownian Motion

There is an analytic solution for both options under the GBM asset price process [8]. Tables 1a,b show the error of the COS formula, which are exponentially converging in N.

4.1.2 Jump-Diffusion Process

In the paper [7], an implicit numerical PDE method is applied for the pricing of a digital put option, when the underlying process follows a jump-diffusion process. The reference price provided was $v(t_0,x_0)=0.387153$. The corresponding results of the COS formula are presented in Table 1c. The error convergence is exponentially in N. However, a larger value for N is required to reach a certain accuracy compared to the results under GBM. The reason for this is that we now deal with a larger integration interval $[a,b]$, so that a larger value for N is needed for accurate series expansions.

4.1.3 Discontinuous Density Function

If we choose the interval $[a,b]$ sufficiently wide, then the local error of the COS formula can be bounded by (see Sect. 3.1)

$$\varepsilon_{COS}(t_0,x|[a,b],N) \approx \frac{b-a}{2}e^{-r\Delta t}\sum_{k=N}^{+\infty}F_k(x)V_k(T). \qquad (43)$$

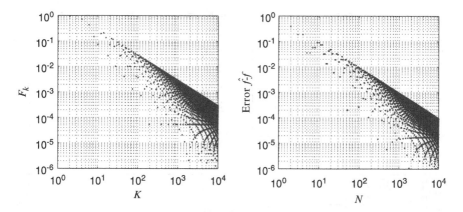

Fig. 2 Coefficients F_k and error of $\hat{f}(y|x_0)$ in log–log scale

For density functions in $C^\infty([a,b])$, the coefficients F_k decay very fast, to be precise, they have an exponential convergence, so that the local error converges exponentially. In this section, we examine a *discontinuous* density function,

$$f(y|x) = x + \frac{1}{2}\, \mathbf{1}_{[y-1,y+1]}, \tag{44}$$

where $\mathbf{1}$ denotes the indicator function, and analyze the convergence of Fourier cosine expansions. The corresponding characteristic function reads

$$\varphi_{\text{levy}}(u) = \frac{e^{iu} - e^{-iu}}{2iu}. \tag{45}$$

Density recovery gives the approximation

$$\hat{f}(y|x) := \sum_{k=0}^{N-1}{}' F_k(x) \cos\left(k\pi \frac{y-a}{b-a}\right). \tag{46}$$

First, we examine the convergence of the coefficients F_k and the error of $\hat{f}(y|x_0)$, for $y = x_0 = 0$. The results in Fig. 2 show that $F_k \sim O(k^{-1})$ and the error converges algebraically with algebraic index of convergence 1.

Next, we price a bull option under the uniform distribution function. The solution is given by

$$v(t_0,x) = e^{-r\Delta t}\frac{1}{2}\int_{x-1}^{x+1} g(e^y)\mathrm{d}y, \tag{47}$$

with $g(.)$ as in (37).

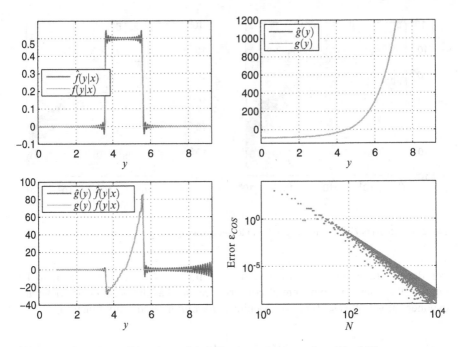

Fig. 3 Density and payoff function and their Fourier cosine expansions ($N = 150$)

We find that $F_k \sim O(k^{-1})$, $V_k \sim O(k^{-2})$, and the option price Error is of order $O(N^{-3})$.

We note that the Fourier cosine series of the discontinuous density function shows a persistent Gibbs phenomenon. The oscillations move towards the discontinuous points $y = x_0 - 1$ and $y = x_0 + 1$ for increasing N, but they do not decrease in magnitude. According to the error analysis in Sect. 3.1 there holds, if $[a,b]$ is sufficiently wide,

$$\hat{v}(t_0, x | [a,b], N) = e^{-r\Delta t} \int_a^b \hat{g}(y) \hat{f}(y|x) \mathrm{d}y. \tag{48}$$

The upper-left plot of Fig. 3 shows the density function and its Fourier-cosine series expansion, with the Gibbs phenomenon. The payoff function is presented in the upper-right plot. Although the slower algebraically convergence of the density, we observe convergence of the option price in N, see lower-right plot. The up-and downward oscillations in the lower-left plot apparently compensate partly.

Table 2 Digital call price under uncertain vol (GBM)

$\mathscr{M}$	100	200	400	800	1600
$\hat{v}(t_0,x_0)$	0.452690	0.449547	0.447311	0.445723	0.444597
$\hat{v}_R(t_0,x_0\|\mathscr{M})$	0.443707	0.443169			

4.2 Uncertain Volatility

In this section, we test the COS method for pricing options under uncertain volatility. We take N sufficiently large, so that convergence of the series approximation is reached. For the control values we use here $\alpha^- = 0.15$, $\alpha^+ = 0.25$.

The numerical method converges in $\mathscr{M}$, the number of time steps, to the solution of the original problem (3). We will use a 4-point Richardson-extrapolation scheme on the option values with small $\mathscr{M}$ to obtain more accurate values. This method is used in [5] to approximate American option values with the help of a few Bermudan option prices. Let $\hat{v}(t_0,x_0|\mathscr{M})$ denote the option value with $\mathscr{M}$ time steps. We calculate the extrapolated value, $\hat{v}_R(t_0,x_0|\mathscr{M})$, by

$$\hat{v}_R(t_0,x_0|\mathscr{M})$$
$$:= \frac{1}{21}\left[64\hat{v}(t_0,x_0|8\mathscr{M}) - 56\hat{v}(t_0,x_0|4\mathscr{M}) + 14\hat{v}(t_0,x_0|2\mathscr{M}) - \hat{v}(t_0,x_0|\mathscr{M})\right].$$
(49)

4.2.1 Geometric Brownian Motion

We start with a digital call option under uncertain volatility. We use the same spatial domain as in [13], namely $[a,b] = [\log 50, \log 160]$, and compare our results.

It is worth mentioning that we do not need to use the extrapolated value $\tilde{V}_k$, with function $\tilde{c}$, from Sect. 3.2. The reason for this is that the value function converges to a constant value if the log-asset price goes to plus or minus infinity. For sufficiently large intervals $[a,b]$, the value function on time lattice t_{m+1} is approximately constant outside the expansion interval. Then, by assuming that N is chosen sufficiently large, the function $\hat{v}(t_{m+1},x)$ is also accurate outside $[a,b]$ and the local error of the COS formula at time t_m is small for all $y \in [a,b]$. Because of this, extrapolation for the continuation value is not necessary.

The results are presented in Table 2, they are very similar to the prices in [13]. Also the extrapolated values $\hat{v}_R(t_0,x_0|\mathscr{M})$ are accurate. The option value, the Delta $(\partial v/\partial S)$ and Gamma $(\partial^2 v/\partial S^2)$ of the option value, and the control domains are shown in Fig. 4.

Next, the bull option under uncertain volatility is valued. Now we use extrapolation to determine accurate values of the continuation values close to both

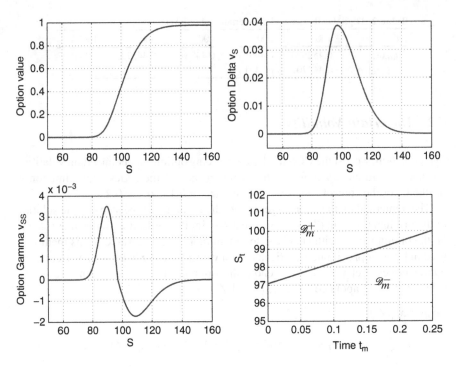

Fig. 4 Digital call under uncertain volatility (GBM, $\mathcal{M} = 1600$)

boundaries a and b. The function $f(y|x,\alpha)$ represents a normal density function with distribution

$$\mathcal{N}\left(x + \left(\mu - \frac{1}{2}\alpha^2\right)\Delta t, \alpha\sqrt{\Delta t}\right). \tag{50}$$

We can presume that the continuation value is well approximated on $[x^{**}, x^*]$, with

$$x^{**} := a - \left(\mu - \frac{1}{2}(\alpha^+)^2\right)\Delta t + 5\alpha^+\sqrt{\Delta t}, \tag{51}$$

$$x^* := b - \left(\mu - \frac{1}{2}(\alpha^+)^2\right)\Delta t - 5\alpha^+\sqrt{\Delta t}. \tag{52}$$

Table 3 presents the option values. Figure 5 shows the option value calculated without and with extrapolation technique. The true value is calculated by using a very large computational domain, so that errors in the vicinity of the boundaries do not affect the option values in the domain $S_0 \in [50, 160]$. Then, a significant larger number of cosine terms is necessary to reach a certain accuracy.

Table 3 Bull under uncertain volatility, with extrapolation, GBM

$\mathscr{M}$	100	200	400	800	1600
$\hat{v}(t_0,x_0)$	0.2342475	0.2313446	0.2298836	0.2291501	0.2287825
$\hat{v}_R(t_0,x_0\|\mathscr{M})$	0.2284146	0.2284142			

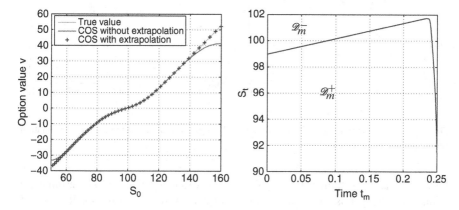

Fig. 5 Bull option under uncertain volatility (GBM)

Table 4 Digital call price under uncertain volatility, jump diffusion

$\mathscr{M}$	100	200	400	800	1600
$\hat{v}(t_0,x_0)$	0.4523049	0.4492045	0.4469982	0.4454318	0.4443220
$\hat{v}_R(t_0,x_0\|\mathscr{M})$	0.4434415	0.4429120			

4.2.2 Merton's Jump-Diffusion Process

We end this section with an option pricing problem where the underlying asset prices is a jump diffusion. Table 4 presents the results for a digital call options. Again the COS method performs highly satisfactorily.

Remark 3 (Convergence in $\mathscr{M}$). The digital call prices converge with order 1/2 in $\mathscr{M}$, see Tables 2 and 4, whereas the bull option converges with order 1 (Table 3). Probably the properties of the payoff function give rise to different convergence rates. However, more research is needed to understand this behavior.

Remark 4 (Incorrect control). In the previous experiments we took N sufficiently large, so that the COS formula for the continuation values was sufficiently accurate. If we would choose N too small, the COS formula may give incorrect control values. This is demonstrated in Fig. 6, where the left-hand side picture shows the continuation values for both control choices with $N = 2^5$, which is apparently too small. The right-hand side picture shows the difference between the two functions

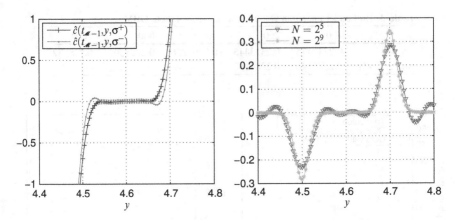

Fig. 6 *Left*: continuation values with small $N = 2^5$, *right*: difference between two continuation values for small $N = 2^5$ and high $N = 2^9$

for $N = 2^5$ and a sufficiently large $N = 2^9$. This function is used to determine the optimal volatility. It is clear that we find incorrect control values for small values of N and the convergence result 2 only holds for sufficiently large N.

5 Conclusion

In this paper, we presented a Fourier method for pricing options under uncertain volatility, based on the models from [11, 13]. The method relies on the dynamic programming principle and the COS formula [4], which is based on Fourier cosine series expansions. A recursive algorithm has been defined, based on the recursive recovery of the series coefficients. With the use of the Fast Fourier Transform algorithm, we achieve a computational complexity of order $O(N \log_2 N)$ per time step, where N denotes the number of terms in the series expansion.

Analysis of the local error enabled us to further improve the method by using an extrapolation method for the area in which the COS formula may give inaccurate continuation values. Extrapolation by Taylor expansion can easily be applied as the derivatives of approximated continuation values can be directly computed based on the COS formula. An exponentially converging error, in N, is found for a sufficiently accurate extrapolation method, $[a, b] \subset \mathbb{R}$ sufficiently wide and a probability density function in the class $C^\infty([a, b])$. A discontinuous density function results in algebraic convergence.

We tested our numerical method by pricing digital and bull options. The COS method for pricing problems under uncertain volatility performed highly satisfactorily.

References

1. Boyd, J.P.: Chebyshev and Fourier Spectral Methods. Courier Dover Publications, New York (2001)
2. Crandall, M.G., Ishii, H., Lions, P.L.: A user's guide to viscosity solutions of second order partial differential equations. Bull. Am. Math. Soc. **27**, 1–67 (1992)
3. Duffie, D., Pan, J., Singleton, K.J.: Transform analysis and asset pricing for affine jump-diffusions. Econometrica **68**(6), 1343–1376 (2000)
4. Fang, F., Oosterlee, C.W.: A novel pricing method for European options based on Fourier-cosine series expansions. SIAM J. Sci. Comput. **31**(2), 826–848 (2008)
5. Fang, F., Oosterlee, C.W.: Pricing early-exercise and discrete barrier options by Fourier-cosine series expansions. Numer. Math. **114**(1), 27–62 (2009)
6. Forsyth, P.A., Labahn, G.: Numerical methods for controlled Hamilton-Jacobi-Bellman PDEs in finance. J. Comput. Finance **11**(2), 1–44 (2007)
7. d'Halluin, Y., Forsyth, P.A., Vetzal, K.R.: Robust numerical methods for contingent claims under jump diffusion processes. IMA J. Numer. Anal. **25**(1), 87–112 (2005)
8. Hull, J.C.: Options, Futures and Other Derivatives. Pearson/Prentice-Hall, New Jersey (2009)
9. Lions, P.L.: Optimal control of diffusion processes and Hamilton-Jacobi-Bellman equations part 2: viscosity solutions and uniqueness. Comm. Part. Differ. Equat. **8**(11), 1229–1276 (1983)
10. Pham, H.: Continuous-Time Stochastic Control and Optimization with Financial Applications. Springer-Verlag, Berlin (2009)
11. Pooley, D.M., Forsyth, P.A., Vetzal, K.R.: Numerical convergence properties of option pricing PDEs with uncertain volatility. IMA J. Numer. Anal. **23**(2), 241–267 (2003)
12. Ruijter, M.J.: On the Fourier cosine series expansion (COS) method for stochastic control problems. Working paper (2011)
13. Zhang, K., Wong, S.: A computational scheme for uncertain volatility model in option pricing. **59**(8), 1754–1767 (2009)

Fast Fourier Transform Option Pricing: Efficient Approximation Methods Under Multi-Factor Stochastic Volatility and Jumps

J.P.F. Charpin and M. Cummins

Abstract Fourier option-pricing methods are popular due to the dual benefits of wide applicability and computational efficiency. The literature tends to focus on a limited subset of models with analytic conditional characteristic functions (CCFs). Models that require numerical solutions of the CCF undermine the efficiency of Fourier methods. To tackle this problem, an ad hoc approximate numerical method was developed that provide CCF values accurately much faster than traditional methods. This approximation, based on averaging, Taylor expansions and asymptotic behaviour of the CCFs, is presented and tested for a range of affine models, with multi-factor stochastic volatility and jumps. The approximation leads to average run-time accelerations up to 50 times those of other numerical implementations, with very low absolute and relative errors reported.

1 Introduction

The application of Fourier transform theory to the pricing of options and general contingent claims is well established in the literature. Carr et al. [7] provide a comprehensive overview. Such methods are popular due to the dual benefits of applicability to a wide class of model dynamics (e.g., affine jump diffusion and Levy processes) and significant computational efficiencies over more traditional option-pricing methods. Heston [15] presents the seminal work on applying the

J.P.F. Charpin (✉)
Department of Mathematics and Statistics, Mathematics Application
Consortium for Science and Industry (MACSI), College of Science and Engineering,
University of Limerick, Limerick, Ireland
e-mail: jean.charpin@ul.ie

M. Cummins
DCU Business School, Dublin City University, Dublin 9, Ireland
e-mail: mark.cummins@dcu.ie

M. Cummins et al. (eds.), *Topics in Numerical Methods for Finance*, Springer Proceedings in Mathematics & Statistics 19, DOI 10.1007/978-1-4614-3433-7_7,
© Springer Science+Business Media New York 2012

Fourier transform, deriving the required conditional characteristic function (CCF) specifications under a two-factor affine stochastic volatility diffusion model. In a range of pricing and empirical applications, some authors [1, 2, 8, 9, 23] consider more generalized affine jump-diffusion models, deriving the required CCFs. Duffie et al. [11] present a generalisation and unification of integral transform pricing under a general multi-factor affine jump-diffusion framework. Carr and Madan [5] show how even greater computational efficiency can be achieved through the application of the fast Fourier transform (FFT), returning option prices for a range of (log-)strike prices simultaneously. Additional FFT applications in exotic and early exercise settings are given by [4, 10, 16–18, 20, 25]. Other authors [19, 21, 22] discuss error control and stability issues.

The flexibility that Fourier and FFT methods offer reflects the ease of transitioning from one underlying model to another through simply altering the CCF specification. Obtaining the CCF under the affine jump-diffusion model framework involves solving the ordinary differential equation (ODE) system detailed by [11], with the Levy–Khintchine representation defining the CCF for Levy processes. This study focuses on the affine jump-diffusion model class. Much of the literature tends to focus on a limited subset of models with analytic CCFs, particularly for empirical applications, so as to maximise the computational efficiencies of the Fourier and FFT methods. The requirement to numerically solve an ODE system for the CCF significantly undermines these efficiencies. This study extends the literature through developing efficient approximations in solving for the CCF, for models with multi-factor stochastic volatility and jumps. A range of jump-augmented versions of the three-factor stochastic volatility diffusion model of [11], which incorporates a stochastic long-run mean volatility level, are considered. The inclusion of a second stochastic volatility component is motivated to potentially capture market volatility better. Each model requires numerical solution of the ODE system that defines the CCF. The approximations developed in this study lead to dramatic reductions in the run-times for FFT option pricing.

Few authors have considered approximation methods for affine models. Glasserman and Kim [14] develop saddlepoint approximations for Fourier transform option pricing that circumvent the need to numerically evaluate the CCF. Whereas the Fourier inversions would require numerical solution of the ODE system at each evaluation point in a numerical integration, saddlepoint approximations require only one ODE system to be solved. However, the dimension of the ODE system depends on the saddlepoint approximation method used and the order of derivatives required. Carr and Madan [6] use saddlepoint approximations but to address the numerical instability of standard Fourier methods for deeply out-of-the-money options. Takahashi and Takehara [24] consider a Fourier transform with an asymptotic expansion approximation for currency option pricing under a very general framework, assuming a jump-diffusion spot exchange rate, mean-reverting diffusion variance and LIBOR market model interest rates. Filipovic et al. [13] present approximate density methods with a view to avoiding Fourier inversion techniques altogether.

The approximation method developed here differs from these studies in working directly on the ODE system for each model and presenting a semi-analytic solution. For the models considered here, in some cases, exact solutions exist in some cases. However, in general, the non-linear and coupled nature of the equations makes this impossible. To tackle the lack of efficiency due to the numerical solution of the CCFs, an ad hoc approximate numerical method was developed. This approximation uses the analytical solutions when possible and otherwise averaging, Taylor expansions and asymptotic behaviour of the CCFs. The method is presented and tested for a range of affine models, with multi-factor stochastic volatility and jumps. The accuracy and speed of the method will also be investigated. This study will be carried out using the standard using the MATLAB *ode45* solver. This function implements a Runge–Kutta method with a variable step size for efficient computation. These numerical results will be used as a benchmark to evaluate the acceleration of the method: MATLAB *ode45* treats all the models used in the present study in the same way and the program will therefore not depend on the model used.

The remainder of the paper is organized as follows. Section 2 provides an overview of FFT option pricing. Section 3 presents the three-factor stochastic volatility diffusion model of [11], in addition to the jump-augmented models with constant and stochastic jump intensities. The semi-analytic approximations are given in Sects. 4 and 5 , along with pricing errors and run-time efficiencies relative to the corresponding numerical implementations. Section 6 concludes.

2   The FFT and Option Pricing

The recent focus on the use of the Fourier transform and the FFT in options pricing is motivated by the significant benefits offered:

1. Applicability to a wide class of model dynamics (e.g., affine jump-diffusion and Levy processes).
2. Impressive computational efficiencies.

More traditional methods fail to combine these two characteristics. In particular, Monte Carlo simulation methods offer significant flexibility in terms of model complexity but are computationally expensive, whereas lattice methods offer significant computational efficiencies but are not easily extended to more complex model dynamics.

Carr and Madan [5] were the first authors to apply the FFT in options pricing. The European-exercise call option price should first be modified appropriately to enforce square integrability, such that the Fourier transform exists. The option price can then be retrieved by means of Fourier inversion of a complex-valued function involving the CCF. If

$$V(k) \equiv e^{-r(T-t)} E_t^{\mathcal{Q}} \left[\max \left(e^{s_T} - e^k, 0 \right) \right]$$

is the value of a European-exercise call option, where the notation is defined in the nomenclature. Carr and Madan [5] show that this value may be retrieved by means of the following Fourier inversion:

$$V(k) = e^{-\alpha^* k} \int_{-\infty}^{\infty} \phi(v) e^{i2\pi v k} dv,$$

where the α^* term is the dampening parameter used in forcing square integrability on the option price, $\alpha^* \in \mathbb{R}^+$, and

$$\phi(v) \equiv \frac{e^{-r(T-t)} \psi^2 \left(i \left[w(v) - (\alpha^* + 1)i\right]\right)}{(\alpha^*)^2 + \alpha^* - w^2(v) + i(2\alpha^* + 1)w(v)};$$

$w(v) \equiv -2\pi v$ and the CCF

$$\psi^2(\tilde{u}) \equiv E_t^2 \left[e^{\tilde{u} \cdot X_T}\right].$$

where $\tilde{u} \in \mathbb{C}$. The Fourier inversion above lends itself to efficient evaluation using an inverse FFT routine. Firstly, the expression is approximated by the following inverse discrete Fourier transform (DFT):

$$\tilde{V}(k_m) = \Delta v \sum_{n=0}^{N-1} \phi(v_n) e^{i2\pi v_n k_m}, \quad m = 0, \ldots, N-1, \tag{1}$$

where N denotes the resolution of the DFT, $k_m \equiv m\Delta k$, Δk is the spacing in the log-strike price domain, $v_n \equiv n\Delta v$ and Δv denotes the spacing in the Fourier domain. Expression (1) therefore simplifies to:

$$\tilde{V}(k_m) = \Delta v \sum_{n=0}^{N-1} \phi(v_n) e^{i2\pi n\Delta v m\Delta k}$$

$$= \Delta v \sum_{n=0}^{N-1} \phi(v_n) e^{i2\pi nm/N}, \quad m = 0, \ldots, N-1. \tag{2}$$

The corresponding call option values can then be obtained directly from these results using the formula:

$$V(k_m) = e^{-\alpha^* k_m} \tilde{V}(k_m), \quad m = 0, \ldots, N-1. \tag{3}$$

Equations (2) and (3) show that the FFT offers an efficient methodology, calculating the price of N option values with a very limited computational cost: the N values $\phi(v_m)$ lead to N different option prices $V(k_m)$.

3 Multi-factor Stochastic Volatility and Jumps

Duffie et al. [11] propose a three factor stochastic volatility diffusion model. For this, the asset price, variance and long-run mean variance are represented as stochastic variables according to the following dynamics:

$$
d \begin{pmatrix} s_t \\ v_t \\ \bar{v}_t \end{pmatrix} = \begin{pmatrix} r - \delta - \frac{1}{2}v_t \\ \kappa(\bar{v}_t - v_t) \\ \kappa_0(\bar{v} - \bar{v}_t) \end{pmatrix} dt + \begin{pmatrix} \sqrt{v_t} & 0 & 0 \\ \sigma\rho\sqrt{v_t} & \sigma\sqrt{1-\rho^2}\sqrt{v_t} & 0 \\ 0 & 0 & \sigma_0\sqrt{\bar{v}_t} \end{pmatrix} dW_t,
$$

where the notation is defined in the nomenclature. Herein, this model is referred to as the SVV model. Using, $\tau \equiv T - t$, the CCF in this case is given by

$$
\psi^{\mathcal{Q}}(\tilde{u}) = e^{\alpha(\tau;\tilde{u}) + \tilde{u}s_t + \beta_1(\tau;\tilde{u})v_t + \beta_2(\tau;\tilde{u})\bar{v}_t}, \tag{4}
$$

where the functions $\beta_1(\tau;\tilde{u})$, $\beta_2(\tau;\tilde{u})$ and $\alpha(\tau;\tilde{u})$ solve the following system of ODEs:

$$
\dot{\beta}_1(\tau) = \frac{1}{2}\sigma^2\beta_1^2(\tau) + (\tilde{u}\sigma\rho - \kappa)\beta_1(\tau) + \frac{1}{2}(\tilde{u}^2 - \tilde{u}) \tag{5}
$$

$$
\dot{\beta}_2(\tau) = \frac{1}{2}\sigma_0^2\beta_2^2(\tau) - \kappa_0\beta_2(\tau) + \kappa\beta_1(\tau) \tag{6}
$$

$$
\dot{\alpha}(\tau) = (r - \delta)\tilde{u} + \kappa_0\bar{v}\beta_2(\tau), \tag{7}
$$

subject to the boundary conditions $\beta_1(0) = \beta_2(0) = \alpha(0) = 0$.

This completes the model. However, jumps are not included in the SVV equations (5)–(7). The governing equations for the functions α, β_1 and β_2 will therefore be slightly modified: in Sect. 3.1, constant jump intensity models are presented while Sect. 3.2 introduces stochastic jump intensities.

3.1 SVV-JD Models: Constant Jump Intensity

The general jump-augmented version of the SVV model in this scenario is given by

$$
d \begin{pmatrix} s_t \\ v_t \\ \bar{v}_t \end{pmatrix} = \begin{pmatrix} r - \delta - \frac{1}{2}v_t - \lambda\xi \\ \kappa(\bar{v}_t - v_t) \\ \kappa_0(\bar{v} - \bar{v}_t) \end{pmatrix} dt + \begin{pmatrix} \sqrt{v_t} & 0 & 0 \\ \sigma\rho\sqrt{v_t} & \sigma\sqrt{1-\rho^2}\sqrt{v_t} & 0 \\ 0 & 0 & \sigma_0\sqrt{\bar{v}_t} \end{pmatrix} dW_t + dZ_t,
$$

where Z_t is a pure jump process in $\mathbb{R}^3$, with constant jump intensity λ. The jump-size distribution has an associated transform of $\Xi(\tilde{u}, \beta_1(\tau), \beta_2(\tau))$, such that $\xi = \Xi(1, 0, 0) - 1$.

In this situation, the three functions $\beta_1(\tau;\tilde{u})$, $\beta_2(\tau;\tilde{u})$ and $\alpha(\tau;\tilde{u})$ solve the following system of ODEs:

$$\dot{\beta}_1(\tau) = \frac{1}{2}\sigma^2\beta_1^2(\tau) + (\tilde{u}\sigma\rho - \kappa)\beta_1(\tau) + \frac{1}{2}(\tilde{u}^2 - \tilde{u})$$

$$\dot{\beta}_2(\tau) = \frac{1}{2}\sigma_0^2\beta_2^2(\tau) - \kappa_0\beta_2(\tau) + \kappa\beta_1(\tau)$$

$$\dot{\alpha}(\tau) = (r - \delta)\tilde{u} + \kappa_0\bar{v}\beta_2(\tau) - \lambda(\xi\tilde{u} + 1) + \lambda\Xi(\tilde{u},\beta_1(\tau),\beta_2(\tau)),$$

subject to the boundary conditions $\beta_1(0) = \beta_2(0) = \alpha(0) = 0$. The only adjustment appears in the ODE for $\alpha(\tau;\tilde{u})$. Three particular jump specifications modelled by $\Xi(\tilde{u},\beta_1(\tau),\beta_2(\tau))$, will be considered as follows—the subscript *CJI* is used to highlight the constant jump intensity feature:

- **SVV-JS$_{\text{CJI}}$:** allows for jumps in the log-asset price (s_t) process with associated jump sizes given by a normal distribution with mean μ_s and variance σ_s^2. For this model, the jump transform is

$$\Xi(\tilde{u},\beta_1(\tau),\beta_2(\tau)) = \Xi(\tilde{u}) = \exp\left(\mu_s\tilde{u} + \frac{1}{2}\sigma_s^2\tilde{u}^2\right),$$

and the integration of this transform function over the life of the option has the following analytic form:

$$\int_0^\tau \Xi(\tilde{u})\,d\vartheta = \exp\left(\mu_s\tilde{u} + \frac{1}{2}\sigma_s^2\tilde{u}^2\right)\tau.$$

- **SVV-JV$_{\text{CJI}}$:** allows for jumps in the variance (v_t) process with associated jump sizes given by an exponential distribution with mean μ_v. For this model, the jump transform is

$$\Xi(\tilde{u},\beta_1(\tau),\beta_2(\tau)) = \Xi(\beta_1(\tau)) = \frac{1}{1 - \mu_v\beta_1(\tau)},$$

and the integration of this transform function over the life of the option has the following analytic form:

$$\int_0^\tau \Xi(\beta_1(\vartheta))\,d\vartheta = \frac{\gamma - \zeta}{\gamma - \zeta + \mu_v\eta}\tau$$

$$-\frac{2\eta\mu_v}{\gamma^2 - (\zeta - \mu_v\eta)^2}\ln\left[1 - \frac{(\gamma + \zeta - \mu_v\eta)(1 - e^{-\gamma\tau})}{2\gamma}\right],$$

where

$$\eta \equiv \tilde{u}(1 - \tilde{u}),$$

$$\zeta \equiv \tilde{u}\rho\sigma_v - \varkappa_v,$$

$$\gamma \equiv \sqrt{\zeta^2 + \eta\sigma_v^2}.$$

- **SVV-JJ$_{CJI}$:** allows for simultaneous and correlated jumps in the log-asset price (s_t) and variance (v_t) processes. The jump sizes in variance are given by an exponential marginal distribution with mean $\mu_{c,v}$. Conditional on a jump in variance of j_v^0, the jump size in the log-asset price is given by a normal distribution with mean $\mu_{c,s} + \rho_J j_v^0$ and variance $\sigma_{c,s}^2$. For this model, the jump transform is

$$\Xi\left(\tilde{u}, \beta_1\left(\tau\right), \beta_2\left(\tau\right)\right) = \Xi\left(\tilde{u}, \beta_1\left(\tau\right)\right) = \frac{\exp\left(\mu_{c,s}\tilde{u} + \frac{1}{2}\sigma_{c,s}^2\tilde{u}^2\right)}{1 - \mu_{c,v}\left(\beta_1\left(\tau\right) + \rho_J\tilde{u}\right)},$$

and the integration of this transform function over the life of the option has the following analytic form:

$$\int_0^\tau \Xi\left(\tilde{u}, \beta_1\left(\vartheta\right)\right) d\vartheta = \exp\left(\mu_{c,s}\tilde{u} + \frac{1}{2}\sigma_{c,s}^2\tilde{u}^2\right) d,$$

where

$$d \equiv \frac{\gamma - \zeta}{c\left(\gamma - \zeta\right) + \mu_{c,v}\eta}\tau$$
$$- \frac{2\eta\mu_{c,v}}{\left(c\gamma\right)^2 - \left(c\zeta - \mu_{c,v}\eta\right)^2} \ln\left[1 - \frac{\left[c(\gamma + \zeta) - \mu_{c,v}\eta\right]\left(1 - e^{-\gamma\tau}\right)}{2c\gamma}\right]$$

and

$$c \equiv 1 - \rho_J\mu_{c,v}\tilde{u}.$$

This completes the constant jump intensity models. The next section considers the stochastic jump intensity models.

3.2 SVV-JD Models: Stochastic Jump Intensity

A natural extension of the models presented in the previous section replaces the constant jump intensity assumption with one of stochastic jump intensity. Intensities are assumed proportional to variance v_t, i.e., the jump intensity is assumed to be given by λv_t, in line with the literature, for instance [3, 12]. This is consistent with the leverage effect, observable in the equity markets in particular, whereby market declines lead to increased volatility. Hence, a stochastic jump intensity that is proportional to variance increases the likelihood of jumps during market distress. The general jump-augmented version of the SVV model in this scenario is given by

$$d\begin{pmatrix} s_t \\ v_t \\ \bar{v}_t \end{pmatrix} = \begin{pmatrix} r - \delta - \left(\frac{1}{2} + \lambda\xi\right)v_t \\ \kappa\left(\bar{v}_t - v_t\right) \\ \kappa_0\left(\bar{v} - \bar{v}_t\right) \end{pmatrix} dt + \begin{pmatrix} \sqrt{v_t} & 0 & 0 \\ \sigma\rho\sqrt{v_t} & \sigma\sqrt{1 - \rho^2}\sqrt{v_t} & 0 \\ 0 & 0 & \sigma_0\sqrt{\bar{v}_t} \end{pmatrix} dW_t + dZ_t,$$

where Z_t is a pure jump process in $\mathbb{R}^3$, with stochastic jump intensity λv_t; the jump-size distribution has an associated transform of $\Xi\left(\tilde{u}, \beta_1\left(\tau\right), \beta_2\left(\tau\right)\right)$, such that $\xi = \Xi\left(1,0,0\right) - 1$.

In these general jump-diffusion models, the functions $\beta_1\left(\tau; \tilde{u}\right)$, $\beta_2\left(\tau; \tilde{u}\right)$ and $\alpha\left(\tau; \tilde{u}\right)$ solve the following system of ODEs:

$$\dot{\beta_1}\left(\tau\right) = \frac{1}{2}\sigma^2\beta_1^2\left(\tau\right) + \left(\tilde{u}\sigma\rho - \kappa\right)\beta_1\left(\tau\right) + \frac{1}{2}\left(\tilde{u}^2 - \tilde{u}\right) - \lambda\left(\xi\tilde{u} + 1\right)$$
$$+ \lambda\Xi\left(\tilde{u}, \beta_1\left(\tau\right), \beta_2\left(\tau\right)\right)$$

$$\dot{\beta_2}\left(\tau\right) = \frac{1}{2}\sigma_0^2\beta_2^2\left(\tau\right) - \kappa_0\beta_2\left(\tau\right) + \kappa\beta_1\left(\tau\right)$$

$$\dot{\alpha}\left(\tau\right) = \left(r - \delta\right)\tilde{u} + \kappa_0\bar{v}\beta_2\left(\tau\right),$$

subject to the boundary conditions $\beta_1\left(0\right) = \beta_2\left(0\right) = \alpha\left(0\right) = 0$.

As in the previous section, three specific jump specifications are assumed and these models are herein referred to as the SVV-JS$_{SJI}$, SVV-JV$_{SJI}$ and SVV-JJ$_{SJI}$ models. For the cases above, the equations governing α, β_1 and β_2 must be solved. As the equations are non-linear and coupled, adequate numerical methods should be used. This will be detailed in the next section.

4   Semi-Analytical Solutions for the SVV Model

The use of FFTs improves computational efficiency dramatically over other option-pricing techniques such as Monte Carlo simulation. However, if numerical solution of the CCF is required then when coding the model using a standard package like MATLAB, the computing time still remains extremely long. In the code, the functions α, β_1 and β_2 defined in the previous section are calculated using the MATLAB *ode45* solver. This in-built function produces very accurate approximations of their values but is extremely time consuming. Since the values $\alpha(\tau)$, $\beta_1(\tau)$ and $\beta_2(\tau)$ are calculated thousands of times when estimating parameters as part of model calibration, this aspect of the computation should be optimised as much as possible.

The SVV model is the simplest of the cases presented in the previous section. As already mentioned, the system of equations governing the CCF is:

$$\dot{\beta_1}\left(\tau\right) = \frac{1}{2}\sigma^2\beta_1^2\left(\tau\right) + \left(\tilde{u}\sigma\rho - \kappa\right)\beta_1\left(\tau\right) + \frac{1}{2}\left(\tilde{u}^2 - \tilde{u}\right) \tag{8}$$

$$\dot{\beta_2}\left(\tau\right) = \frac{1}{2}\sigma_0^2\beta_2^2\left(\tau\right) - \kappa_0\beta_2\left(\tau\right) + \kappa\beta_1\left(\tau\right) \tag{9}$$

$$\dot{\alpha}\left(\tau\right) = \left(r - \delta\right)\tilde{u} + \kappa_0\bar{v}\beta_2\left(\tau\right), \tag{10}$$

and the system is solved subject to the following initial conditions: $\beta_1(0) = \beta_2(0) = \alpha(0) = 0$. This system involves a mixture of three linear/non-linear equations, coupled or uncoupled. The first equation is non-linear but uncoupled with the other two. An analytical solution exists for this equation and the corresponding results will be presented in Sect. 4.1. The second equation is non-linear and coupled with the first equation. This is the most difficult equation of the system and solutions for this aspect of the problem are presented in Sect. 4.2. Finally, the third equation appears as an integral of the solution of the second equation. Solutions for this last equation will be detailed in Sect. 4.3. The results obtained with this semi-analytical method will then be compared with the *ode45* solutions and the run-time efficiencies of the approximations will be evaluated in Sect. 4.4.

4.1 Analytical Solution for β_1

Equation (8) admits an explicit analytical solution. None of the parameters depend on time τ. Using standard integration techniques, the solution of (8) is:

$$\beta_1(\tau) = z_1 z_2 \frac{e^{a(z_1 - z_2)\tau} - 1}{z_1 e^{a(z_1 - z_2)\tau} - z_2}, \tag{11}$$

where $a = \sigma^2/2$ and z_1 and z_2 are the roots of

$$\frac{1}{2}\sigma^2 z^2 + (\tilde{u}\sigma\rho - \kappa)z + \frac{1}{2}(\tilde{u}^2 - \tilde{u}) = 0,$$

with $\text{Real}(z_1) > \text{Real}(z_2)$.

The equation will have to be solved numerous times as the parameter $\tilde{u}$ varies with the maturity and moneyness of the option. Figure 1 shows the real and imaginary parts of function β_1 for three different cases. In Fig. 1a, the real and imaginary parts behave nearly linearly. For longer times, the function reaches its asymptotic value very quickly as is shown on Fig. 1c. For intermediate cases, the curves first behave linearly and evolve towards the asymptotic value without reaching it. As β_1 appears directly in (9), these different behaviours will directly affect the solution method for β_2. This will now be detailed.

4.2 Analytical Solution for β_2

Equations (8) and (9) have very similar structures. As β_1 appears explicitly in (9), the analytical solution for β_2 will reflect the different behaviours of β_1. Three regions will be considered for β_2:

1. For values of $\tau < \tau_{\text{ref}}$, where τ_{ref} depends of $\tilde{u}$ but remains undefined for the moment, Taylor series will lead to a very accurate solutions.

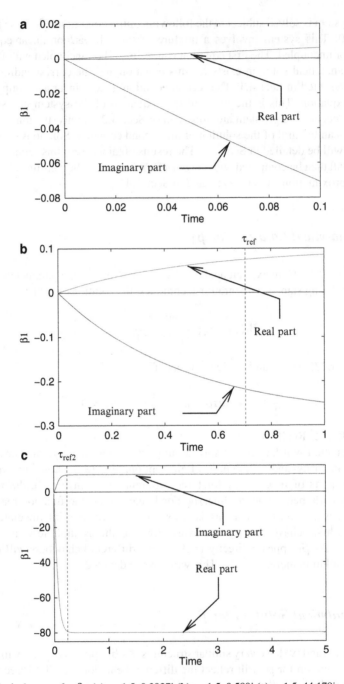

Fig. 1 Typical curves for β_1. (**a**) u=1.5–0.3927i (**b**) u=1.5–0.589i (**c**) u=1.5–44.178i

2. For $\tau_{\mathrm{ref}} < \tau < \tau_{\mathrm{ref}_2}$, β_1 will be replaced by its average over a subsection of the interval. With appropriate boundary conditions, an approximate solution will be calculated. Here again, τ_{ref_2} still needs to be determined but it will depend on the maturity of the function through $\tilde{u}$. As $\beta_1(0) = 0$, this method would lead to poor results close to $\tau = 0$ and this explains why the Taylor series are necessary for $\tau < \tau_{\mathrm{ref}}$.

3. For $\tau > \tau_{\mathrm{ref}_2}$, the function β_1 will be replaced by its asymptotic value and the method presented in the previous point remains valid.

4.2.1 Taylor Series

For small values of τ, Taylor expansions of the functions β_1 and β_2 are:

$$\beta_1(\tau) = a_1\tau + a_2\tau^2 + a_3\tau^3 + a_4\tau^4 + a_5\tau^5 + a_6\tau^6 + a_7\tau^7 + a_8\tau^8 + a_9\tau^9$$
$$\beta_2(\tau) = b_1\tau + b_2\tau^2 + b_3\tau^3 + b_4\tau^4 + b_5\tau^5 + b_6\tau^6 + b_7\tau^7 + b_8\tau^8 + b_9\tau^9$$

where

$$a_1 = az_1z_2;$$

$$a_2 = \frac{a^2}{2}z_1z_2(z_1 + z_2)$$

$$a_3 = \frac{a^3}{6}z_1z_2\left(z_1^2 + 4z_1z_2 + z_2^2\right)$$

$$a_4 = \frac{a^4}{24}z_1z_2\left(z_1^3 + 11z_1^2z_2 + 11z_1z_2^2 + z_2^3\right)$$

$$a_5 = \frac{a^5}{120}z_1z_2\left(z_1^4 + 26z_1^3z_2 + 66z_1^2z_2^2 + 26z_1z_2^3 + z_2^4\right)$$

$$a_6 = \frac{a^6}{720}z_1z_2\left(z_1^5 + 57z_1^4z_2 + 302z_1^3z_2^2 + 302z_1^2z_2^3 + 57z_1z_2^4 + z_2^5\right)$$

$$a_7 = \frac{a^7}{5040}z_1z_2\left(z_1^6 + 120z_1^5z_2 + 1191z_1^4z_2^2 + 2416z_1^3z_2^3 + 1191z_1^2z_2^4 + 120z_1z_2^5 + z_2^6\right)$$

$$a_8 = \frac{a^8}{40320}z_1z_2\left(z_1^7 + 247z_1^6z_2 + 4293z_1^5z_2^2 + 15619z_1^4z_2^3 + 15619z_1^3z_2^4 + 4293z_1^2z_2^5 \right.$$
$$\left. + 247z_1z_2^6 + z_2^7\right)$$

$$a_9 = \frac{a^9}{362880}z_1z_2\left(z_1^8 + 502z_1^7z_2 + 14608z_1^6z_2^2 + 88234z_1^5z_2^3 + 156190z_1^4z_2^4 \right.$$
$$\left. + 88234z_1^3z_2^5 + 12608z_1^2z_2^6 + 502z_1z_2^7 + z_2^8\right)$$

and

$$b_1 = 0$$

$$b_2 = \frac{\kappa a_1}{2}$$

$$b_3 = \frac{\kappa a_2 - \kappa_0 b_2}{3}$$

$$b_4 = \frac{\kappa a_3 - \kappa_0 b_3}{4}$$

$$b_5 = \frac{\kappa a_4 - \kappa_0 b_4 + a_0 b_2^2}{5}$$

$$b_6 = \frac{\kappa a_5 - \kappa_0 b_5 + 2a_0 b_2 b_3}{6}$$

$$b_7 = \frac{\kappa a_6 - \kappa_0 b_6 + a_0 b_3^2 + 2a_0 b_2 b_4}{7}$$

$$b_8 = \frac{\kappa a_7 - \kappa_0 b_7 + 2a_0 b_2 b_5 + 2a_0 b_3 b_4}{8}$$

$$b_9 = \frac{\kappa a_8 - \kappa_0 b_8 + a_0 b_4^2 + 2a_0 b_3 b_5 + 2a_0 b_2 b_6}{9}$$

These expansions are valid while the β_1 expansion holds, experimentally, this is the case until the exponential argument in (11) is smaller than 1.5. The expansions will therefore hold for

$$\tau < \tau_{\text{ref}} \approx \frac{1.5}{a \text{ real} (z_1 - z_2)}.$$

4.2.2   Averaged Values of β_1

For larger values of τ, (9) is simplified by replacing β_1 by its average value. The equation then writes

$$\dot{\beta}_2 (\tau) = \frac{1}{2} \sigma_0^2 \beta_2^2 (\tau) - \kappa_0 \beta_2 (\tau) + \kappa \bar{\beta}_1.$$

Following the method used in the previous section, an approximate solution for β_2 is

$$\beta_2(\tau) = z_3 z_4 \frac{e^{a_0 (z_3 - z_4) \tau} - 1}{z_3 e^{a_0 (z_3 - z_4) \tau} - z_4} \tag{12}$$

where z_3 and z_4 are the solutions of

$$\frac{1}{2}\sigma_0^2 z^2 - \kappa_0 z + \kappa\bar{\beta}_1 = 0, \tag{13}$$

with $\text{real}(z_3) > \text{real}(z_4)$. The solution will be complete when the average value $\bar{\beta}_1$ is determined. A straightforward integration of equation (11) leads to:

$$\bar{\beta}_1 = \int_{\tau_i}^{\tau_{i+1}} \beta_1(\vartheta)d\vartheta = z_1 - \frac{1}{a(\tau_{i+1}-\tau_i)}\log\left(\frac{z_1 e^{a(z_2-z_1)\tau_{i+1}} - z_2}{z_1 e^{a(z_2-z_1)\tau_i} - z_2}\right).$$

Alternatively, β_1 reaches $\theta = 99.9\%$ of its asymptotic value for

$$\tau > \tau_{\text{ref}_2} \approx \frac{2}{\sigma^2(z_1-z_2)}\log\left(\frac{1-\theta\,|z_2/z_1|}{1-\kappa}\right).$$

In this case, the average value of β_1 is

$$\bar{\beta}_1 = z_1.$$

The interval $[\tau_{\text{ref}}\ \tau]$ is then divided into n_0 intervals if $\tau \le \tau_{\text{ref2}}$. This determines the values for τ_i used in the previous section:

$$\tau_1 = \tau_{\text{ref}}, \qquad \tau_i = \tau_{\text{ref}} + \frac{i-1}{n_0}(\tau-\tau_{\text{ref}}), \quad 2 \le i \le n_0 + 1$$

If $\tau \ge \tau_{\text{ref2}}$, an (n_0+1)th interval $[\tau_{\text{ref}_2}\ \tau]$ is added. Numerical results show that for the range of maturities considered here, $n_0 = 4$ leads to accurate results without compromising the computational time. On each interval, β_1 is replaced by its average or asymptotic value on the interval and an approximated solution is calculated using a method similar to the previous section:

$$\beta_2(\tau) = \frac{z_{3_{n_0-1}}\left(\beta_{2_{n_0-1}} - z_{4_{n_0-1}}\right)}{\beta_{2_{n_0-1}} - z_{4_{n_0-1}} - \left(\beta_{2_{n_0-1}} - z_{3_{n_0-1}}\right)e^{\sigma_0^2\left(z_{3_{n_0-1}} - z_{4_{n_0-1}}\right)(\tau-\tau_{n_0-1})/2}}$$
$$- \frac{z_{4_{n_0-1}}\left(\beta_{2_{n_0-1}} - z_{3_{n_0-1}}\right)e^{\sigma_0^2\left(z_{3_{n_0-1}} - z_{4_{n_0-1}}\right)(\tau-\tau_{n_0-1})/2}}{\beta_{2_{n_0-1}} - z_{4_{n_0-1}} - \left(\beta_{2_{n_0-1}} - z_{3_{n_0-1}}\right)e^{\sigma_0^2\left(z_{3_{n_0-1}} - z_{4_{n_0-1}}\right)(\tau-\tau_{n_0-1})/2}},$$

where the constants z_{3_i} and z_{4_i} are the roots of

$$\frac{1}{2}\sigma_0^2 z^2 - \kappa_0 z + \kappa\bar{\beta}_{1_i} = 0$$

with $\text{Real}(z_{3_i}) > \text{Real}(z_{4_i})$ and

$$\beta_{2_i} = \frac{z_{3_{i-1}}\left(\beta_{2_{i-1}} - z_{4_{i-1}}\right) - z_{4_{i-1}}\left(\beta_{2_{i-1}} - z_{3_{i-1}}\right) e^{\sigma_0^2\left(z_{3_{i-1}} - z_{4_{i-1}}\right)(\tau_i - \tau_{i-1})/2}}{\beta_{2_{i-1}} - z_{4_{i-1}} - \left(\beta_{2_{i-1}} - z_{3_{i-1}}\right) e^{\sigma_0^2\left(z_{3_{i-1}} - z_{4_{i-1}}\right)(\tau_i - \tau_{i-1})/2}},$$

$$\beta_{2_1} = \sum_{i=1}^{9} b_i \tau_{\text{ref}}^i.$$

Figure 2 shows the real and imaginary parts of β_2 for the various cases described above. As could be expected for small values of τ, the approximation of β_2 is extremely accurate as can be seen on Fig. 2a. Figure 2b, c show the results for larger values of τ. The values of the two parameters τ_{ref} and τ_{ref_2} decrease when $\tilde{u}$ increases. For the largest values, as shown on Fig. 2c, the interval $[\tau_{\text{ref}_2}\tau]$ is the largest. The first phases are, however, necessary to guarantee the accuracy of the approximation.

4.3   Semi-Analytical Solution for α

The semi-analytical solution for α reflects the methods used to calculate β_2 as the parameter α is a linear function of β_2. Using the same notation as in the previous sections, the approximation of α is:

- $\tau < \tau_{\text{ref}}$:

$$\alpha(\tau) = (r - \delta)\tilde{u}\tau + \kappa_0\bar{v}\sum_{i=1}^{9} b_i \frac{\tau_{\text{ref}}^{i+1}}{i+1} + \kappa_0\bar{v}\sum_{i=0}^{n-1}$$

$$\times \left[z_{3_i}(\tau_{i+1} - \tau_i) - \frac{2}{\sigma_0^2} \log\left(\frac{\beta_i - z_{4_i} + (z_{3_i} - \beta_i) e^{\sigma_0^2\left(z_{3_i} - z_{4_i}\right)(\tau_{i+1} - \tau_i)/2}}{z_{3_i} - z_{4_i}} \right) \right].$$

- $\tau > \tau_{\text{ref}}$:

$$\alpha(\tau) = (r - \delta)\tilde{u}\tau + \kappa_0\bar{v}\sum_{i=1}^{9} b_i \frac{\tau_{\text{ref}}^{i+1}}{i+1} + \kappa_0\bar{v}\sum_{i=0}^{n}$$

$$\times \left[z_{3_i}(\tau_{i+1} - \tau_i) - \frac{2}{\sigma_0^2} \log\left(\frac{\beta_i - z_{4_i} + (z_{3_i} - \beta_i) e^{\sigma_0^2\left(z_{3_i} - z_{4_i}\right)(\tau_{i+1} - \tau_i)/2}}{z_{3_i} - z_{4_i}} \right) \right].$$

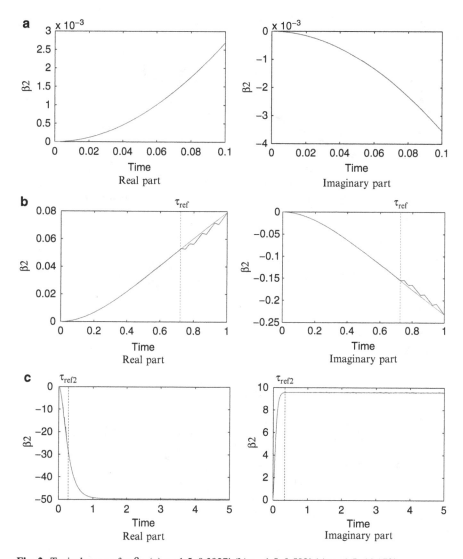

Fig. 2 Typical curves for β_2. (**a**) u=1.5–0.3927i (**b**) u=1.5–0.589i (**c**) u=1.5–44.178i

4.4 Comparison Between the Numerical and Semi-Analytical Solutions

In the previous sections, α, β_1 and β_2 were calculated using approximations leading to reasonably accurate results. However, these three functions appear in the code through more complex computational procedures, see (4) for example. Consequently, the validity of the approximations in the present context can only be evaluated on test cases using the complete optimisation procedure.

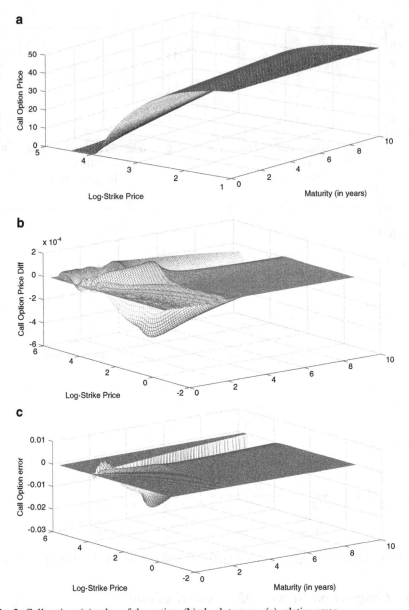

Fig. 3 Call option: (**a**) value of the option, (**b**) absolute error, (**c**) relative error

The approximations are therefore tested for put and call options on a wide range of strike prices (defined by the resolution of the FFT) and maturity times ranging from 3 days to 10 years. The semi-analytical method is benchmarked against the results obtained using the MATLAB *ode45* solver.

Figures 3 and 4 show the difference between the semi-analytical approximation and MATLAB *ode45* solver. In all cases, the error remains very small: the absolute error remains small in all cases, typically less than 5.10^{-4} in absolute terms and 0.02% in relative terms. The approximate analytical solution will therefore be used to replace the results of the MATLAB *ode45* solver as the two methods provide equivalent results. However, the approximate solution is calculated much faster as shown on Fig. 5. The solution of the differential equation is central to the option-pricing algorithm. For the numerical solution method, *ode45* is applied at each point of the FFT resolution. So the overall efficiency of the full option-pricing algorithm is far more important than just the numerical solution part. The code was then run twice:

- In the first instance, the method developed above was used and computation time was measured for a range maturation times.
- In the second instance, *ode45* was used and computation time was measured for the same maturation times.

Computing time for the approximate solution is almost constant at $t \approx 0.6 - 0.8$s while MATLAB *ode45* solver takes up to 1 min to calculate the solutions. Acceleration is then defined as the ratio between these two values. It ranges from about 16 to 80. As the values $\alpha(\tau)$, $\beta_1(\tau)$ and $\beta_2(\tau)$ are calculated thousands of times, the approximate analytical solution will reduce computation times significantly. To complete this study, this semi-analytical solution will be extended to jump models. This will now be performed.

5 Semi-Analytical Solutions: Extension of the Method to the Jump Models

The method presented above has to be modified before it can be applied to the models presented in Sect. 3. For these six models, the equations governing the three functions α, β_1 and β_2 include an additional term in the equation for α or the equation for β_2. For the constant jump intensity models, the numerical method can be applied as such, the integral of the jump transform should only be added to the results for α presented in Sect. 4.3.

The situation is slightly more difficult for the stochastic jump intensity models. For the SVV-JS$_{SJI}$ model, the numerical method can be applied if the parameters z_1 and z_2 are solutions of the equation:

$$\frac{1}{2}\sigma^2 z^2 + (\tilde{u}\sigma\rho - \kappa)z + \frac{1}{2}\left(\tilde{u}^2 - \tilde{u}\right) - \lambda\left(\xi\tilde{u} + 1\right) + \lambda\exp\left(\mu_s\tilde{u} + \frac{1}{2}\sigma_s^2\tilde{u}^2\right) = 0$$

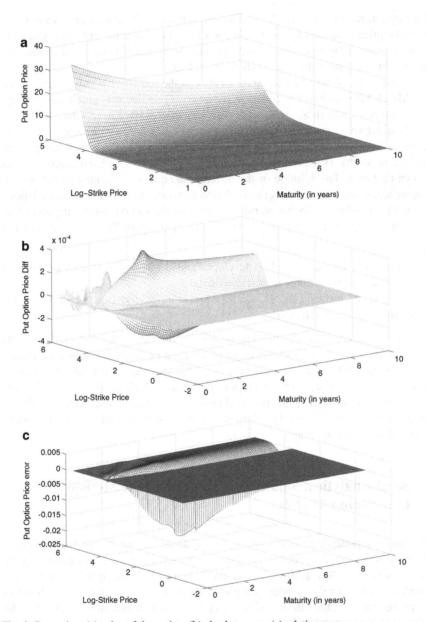

Fig. 4 Put option: (**a**) value of the option, (**b**) absolute error, (**c**) relative error

For the SVV-JJ$_{SJI}$ and SVV-JV$_{SJI}$ jump models, the additional term in the equation governing β_1, $\Xi(\beta_1, \bar{u})$, explicitly depends on β_1. In these two cases, there is no analytical solution for the equation. In this situation, the interval $[0\ \tau]$ is divided into n intervals $[\tau_{i-1}\ \tau_i]$, $1 \leq i \leq n$. On each interval $\Xi(\beta_1, \bar{u})$ is replaced by $\Xi(\bar{\beta}_{1_i}, \bar{u})$

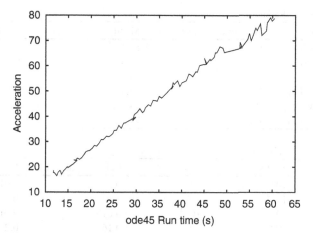

Fig. 5 Acceleration

where $\bar{\beta}_{1_i}$ is the average value of β_1 on an interval $[\tau_{i-1}\ \tau_i]$, still to be determined. This value is calculated using the following algorithm:

1. Initialise $\bar{\beta}_{1_i} = 0$
2. Calculate $z_{1_{i-1}}$ and $z_{2_{i-1}}$ the two roots of the quadratic equation

$$\frac{1}{2}\sigma^2 z^2 + (\tilde{u}\sigma\rho - \kappa)z + \frac{1}{2}\left(\tilde{u}^2 - \tilde{u}\right) + \Xi(\bar{\beta}_{1_i}, \tilde{u}) = 0,$$

where $z_{1_{i-1}}$ and $z_{2_{i-1}}$ are the values of z_1 and z_2 on the interval $[\tau_{i-1}\ \tau_i]$.
3. Calculate the corresponding average value

$$\bar{\beta}_{1_i} = z_{1_{i-1}} - \frac{2}{\sigma^2(\tau_i - \tau_{i-1})}\log\left(\frac{z_{1_{i-1}}e^{\sigma^2\left(z_{2_{i-1}}-z_{1_{i-1}}\right)\tau_i/2} - z_{2_{i-1}}}{z_{1_{i-1}}e^{\sigma^2\left(z_{2_{i-1}}-z_{1_{i-1}}\right)\tau_{i-1}/2} - z_{2_{i-1}}}\right).$$

4. Repeat steps (2) and (3) until convergence is achieved. In all the tests performed, convergence is achieved after 3 or 4 iterations.

As in the previous section, the semi-analytical method is tested for put and call options on a wide range of strike prices and maturity times and is benchmarked against the results obtained using the MATLAB *ode45* solver. Table 1 reports the computational efficiencies achieved under the approximations and Table 2 shows the differences between the semi-analytical and numerical methods. All approximation results are calculated with $n_0 = 4$.

For all cases, the semi-analytical method proves significantly faster than the MATLAB *ode45* function; results are on average computed 45–50 times faster for most jump models. For the SVV-JV$_{\text{SJI}}$ and SVV-JJ$_{\text{SJI}}$ jump models, the acceleration is about four times less. These are the only two models requiring the iterative part of the model as there is no analytical solution for $\beta_1(\tau)$. The iterative algorithm must be repeated on each of the intervals $[\tau_i\ \tau_{i+1}]$ and this causes the lower acceleration.

Table 1 Summary of accelerations

Model	Mininimum acceleration	Maximum acceleration	Average
SVV-JS$_{CJI}$	17.4	81.1	46.5
SVV-JS$_{SJI}$	18	84.4	48
SVV-JV$_{CJI}$	16.9	84.4	46.9
SVV-JV$_{SJI}$	3.5	15.3	9
SVV-JJ$_{CJI}$	18.1	84.5	48.1
SVV-JJ$_{SJI}$	4.8	21.1	12.3

Table 2 Summary of errors

Model	Abs. diff. put	Abs. diff. call	Rel. diff. put (%)	Rel. diff. call (%)
SVV-JS$_{CJI}$	3.10^{-4}	4.10^{-4}	0.008	0.015
SVV-JS$_{SJI}$	5.10^{-4}	5.10^{-4}	0.025	0.02
SVV-JV$_{CJI}$	3.10^{-4}	5.10^{-4}	0.016	0.017
SVV-JV$_{SJI}$	3.10^{-3}	3.10^{-3}	0.16	0.2
SVV-JJ$_{CJI}$	2.10^{-4}	4.10^{-4}	0.007	0.016
SVV-JJ$_{SJI}$	10^{-2}	$2.5.10^{-3}$	0.18	0.13

The acceleration values are averaged for maturities ranging from 3 days to 10 years. For all cases, the acceleration is smaller than average for short maturity times and larger for long maturity times. As in the previous section, the computing time remains approximately constant with the semi-analytic method while it varies significantly when *ode45* is used. This explains again the variation in acceleration.

The differences between the two methods are extremely small in all cases, see Table 2, typically around 10^{-4}–10^{-3} (absolute errors) and less than 0.03% (relative errors) in most cases. For all models, the largest errors occur for strike prices or maturity values, which are not of practical relevance. Here again, the SVV-JV$_{SJI}$ and SVV-JJ$_{SJI}$ jump models provide the least accurate results with differences reaching 0.1%. This is once more related to the difficulties with the approximation of $\beta_1(\tau)$: it is not possible to calculate an exact value of this parameter and the error is transferred to the other two values $\beta_2(\tau)$ and $\alpha(\tau)$ and the errors pile up. However, the resulting differences remain within very acceptable levels (Table 3).

6   Conclusion

Efficient approximations are developed that allow FFT option-pricing methods to be applied to models with multi-factor stochastic volatility and jumps, circumventing the need to numerically solve for the associated CCFs. The approach taken is distinct from other approximation methods in the literature through working directly on the ODE system that defines the CCF and deriving semi-analytic approximate solutions. The combination of average values, Taylor expansions, and asymptotic analysis proves extremely efficient: for the majority of models considered, the

Table 3 Nomenclature

Symbol	Parameter
k	Log-strike price
r	Constant risk-free rate of interest
t	Current time
s_t	Log asset price at time t
v_t	Variance state variable at time t
$\bar{v}_t$	Long-run mean variance state variable at time t
$\bar{v}$	Long-run mean of the long-run mean variance state variable
N	Resolution of the FFT
$\mathcal{Q}$	Risk-neutral probability measure
T	Option maturity
V	Option value
X_t	General state variable process at time t
α^*	Dampening parameter
δ	Continuous dividend yield
κ	Speed of mean reversion of variance
κ_0	Speed of mean reversion of long-run mean variance
λ	Jump intensity parameter
μ	Mean jump size
ρ	Correlation between log asset price and variance
ρ_J	Correlation between simultaneous jumps in asset price and variance
σ	Volatility of variance
σ_0	Volatility of long-run variance
σ	Standard deviation of jump size

approximations report average run-time accelerations of between 45 and 50 times those of the numerical implementations, with very low absolute and relative errors between the opposing methods.

Future research will focus on improving the approximations developed. For the stochastic jump intensity SVV-JS$_{SJI}$ and SVV-JJ$_{SJI}$ models, the run-time accelerations, although impressive, are not as significant as for the other models. Addressing this, in particular, will add significantly to the approximation method. Application to higher dimensional problems and generalisation of the approach will also be pursued.

Acknowledgements Jean Charpin acknowledges the support of the Mathematics Applications Consortium for Science and Industry (www.macsi.ul.ie) funded by the Science Foundation Ireland Mathematics Initiative grant 06/MI/005.

References

1. Bakshi, G., Cao, C., Chen, Z.: Empirical performance of alternative option pricing models. J. Financ. **52**, 2003–2049 (1997)
2. Bates, D.: Jumps and stochastic volatility: exchange rate processes implicit in deutsche mark options. Rev. Financ. Stud. **9**, 69–107 (1996)
3. Bates, D.: Post-'87 crash fears in the S&P 500 futures option market. J. Econometrics. **94**, 181–238 (2000)
4. Benhamou, E.: Fast Fourier transform for discrete Asian options. J. Comput. Financ. **6**, 49–61 (2002)
5. Carr, P., Madan, D.: Option valuation using the fast Fourier transform. J. Comput. Financ. **3**, 463–520 (1999)
6. Carr, P., Madan, D.: Saddlepoint methods for option pricing. J. Comput. Financ. **13**, 49–61 (2009)
7. Carr, P., Géman, H., Madan D., et al.: Option pricing using integral transforms. New York University. Available online. http://www.math.nyu.edu/research/carrp/papers/pdf/integtransform.pdf (2011). Last time accessed May 2012
8. Chacko, G., Das, S.R.: Pricing interest rate derivatives: a general approach. Rev. Financ. Stud. **15**, 195–241 (2002)
9. Das, S.R., Foresi, S.: Exact solutions for bond and option prices with systematic jump risk. Rev. Derivatives Res. **1**, 7–24 (1996)
10. Dempster, M.A.H., Hong, S.S.G.: Spread option valuation and the fast Fourier transform. In: German, H., Madan, D., Pliska, S.R., Vorst, T. (ed.) Mathematical Finance – Bachelier Congress 2000: Selected Papers form the World Congress of the Bachelier Finance Society. Springer-Verlag, Berlin (2001)
11. Duffie, D., Pan, J., Singleton, K.: Transform analysis and option pricing for affine jump-diffusions. Econometrica **68**, 1343–1376 (2000)
12. Eraker, B.: Do stock prices and volatility jump? Reconciling evidence from spot and option prices. J. Financ. **59**, 1367–1404 (2004)
13. Filipovic, D., Mayerhoffer, E., Schneider, P.: Density approximations for multivariate affine jump-diffusion processes [Working Paper]. Available via arXiv. http://arxiv.org/abs/1104.5326 (2011). Last time accessed May 2012
14. Glasserman, P., Kim, K-K.: Saddlepoint approximations for affine jump-diffusion models. J. Econ. Dyn. Control. **33**, 15–36 (2009)
15. Heston, S.L.: A closed-form solution for options with stochastic volatility and applications to bond and currency options. Rev. Financ. Stud. **6**, 327–343 (1993)
16. Hurd, T.R., Zhou, Z.: A Fourier transform method for spread option pricing [Working Paper]. McMaster University. Available via arXiv. http://arxiv.org/PS_cache/arxiv/pdf/0902/0902.3643v1.pdf (2009). Last time accessed May 2012
17. Jackson, K.R., Jaimungal, S., Surkov, V.: Fourier space time-stepping for option pricing with Levy models. J. Comput. Financ. **12**, 1–29 (2008)
18. Jaimungal, S., Surkov, V.: Fourier space time-stepping for option pricing with Levy models [Working Paper]. University of Toronto. Available via Social Science Research Network. http://papers.ssrn.com/sol3/papers.cfm?abstract_id=1302887 (2009). Last time accessed May 2012
19. Lee, R.W.: Option pricing by transform methods: Extensions, unification, and error control. J. Comput. Financ. **7**, 51–86 (2004)
20. Lord, R., Fang, F., Bervoets, F., et al.: A fast and accurate FFT-based method for pricing early-exercise options under Levy processes. SIAM J. Sci. Comput. **30**, 1678–1705 (2008)
21. Lord, R., Kahl, C.: Optimal Fourier inversion in semi-analytical option pricing. J. Comput. Financ. **10**, 1–30 (2007)
22. Minenna, M., Verzella, P.: A revisited and stable Fourier transform method for affine jump diffusion models. J. Bank. Financ. **32**, 2064–2075 (2008)

23. Scott, L.O.: Pricing stock options in a jump-diffusion model with stochastic volatility and interest rates: application of Fourier inversion methods. Math. Financ. **7**, 345–358 (1997)
24. Takahashi, A., Takehara, K.: Fourier transform method with an asymptotic expansion approach: An application to currency options [Working Paper]. University of Tokyo. Available online. http://www.cirje.e.u-tokyo.ac.jp/research/dp/2008/2008cf538.pdf (2009). Last time accessed May 2012
25. Zhylyevskyy, O.: A fast Fourier transform technique for pricing American options under stochastic volatility. Rev. Derivatives Res. **13**, 1–24 (2010)

Pricing Credit Derivatives in a Wiener–Hopf Framework

Daniele Marazzina[†], Gianluca Fusai, and Guido Germano

Abstract We present fast and accurate pricing techniques for credit-derivative contracts when discrete monitoring is applied and the underlying evolves according to an exponential Lévy process. Our pricing approaches are based on the Wiener–Hopf factorization, and their computational cost is independent of the number of monitoring dates. Numerical results are presented in order to validate the pricing algorithm. Moreover, an analysis on the sensitivity of the probability of default and the credit spread term structures with respect to the process parameters is considered.

1 Introduction

In this article, we present fast and accurate pricing techniques for credit derivatives, like defaultable coupon bonds, assuming that default is monitored only on discrete dates and the underlying evolves according to an exponential Lévy process.

[†](The author was partially supported by GNCS-INDAM.)

D. Marazzina (✉)
Department of Mathematics F. Brioschi, Politecnico di Milano, Via E. Bonardi 9,
20133 Milano, Italy
e-mail: daniele.marazzina@polimi.it

G. Fusai
Department of Economics and Business Studies (DiSEI), Università degli Studi del Piemonte Orientale A. Avogadro, Via E. Perrone 18, 28100 Novara, Italy

Department of Finance, Cass Business School, 106 Bunhill Row, EC1Y 8TZ London, UK
e-mail: gianluca.fusai@unipmn.it

G. Germano
Dipartimento di Economics and Business Studies (DiSEI), Università degli Studi del Piemonte Orientale A. Avogadro, Via E. Perrone 18, 28100 Novara, Italy
e-mail: guido.germano@unipmn.it

M. Cummins et al. (eds.), *Topics in Numerical Methods for Finance*, Springer Proceedings in Mathematics & Statistics 19, DOI 10.1007/978-1-4614-3433-7_8,
© Springer Science+Business Media New York 2012

Our pricing approaches, which are related to the maturity randomization algorithm based on the Z-transform presented by Fusai et al. [11, 12], change the problem of computing the default probability into a Wiener–Hopf (WH) factorization problem. This can be solved numerically exploiting the fast Fourier transform (FFT) algorithm. Moreover, the computational cost of our procedures is independent of the number of monitoring dates. Indeed, exploiting the Euler summation, which is a convergence-acceleration technique for alternating series, we can speed up the inverse Z-transform when the probability of default is computed for a large number of monitoring dates.

The connection between pricing problems under Lévy processes and the WH factorization is well known; see, for example, Kyprianou [20] and references therein. Usually, WH methods are applied to the continuous-monitoring case [16–19], assuming an analytical Fourier transform. In [17–19], the authors make particularly stringent assumptions on the form of the characteristic function of the considered Lévy processes. Thanks to this assumption, they provide analytical or semi-analytical formulas for the WH factors. For general characteristic functions, and thus general Lévy processes, fast numerical methods have been proposed, for example, by Kudryavtsev and Levendorskiĭ [16]. Instead, explicit formulas for the WH factors in the discrete monitoring case have been computed only when the underlying asset evolves according to a geometric Brownian motion [3, 13]. Our work, is mainly based on the latter framework, but it extends it substantially considering general Lévy processes.

The main application of the WH technique considered here is to the structural approach: the credit event is defined as the first time that the firm value, i.e., the underlying, drops below a predefined lower barrier. Thus, our framework is similar to that of [6, 9, 10]. Feng and Linetsky [10] present an accurate method based on the Hilbert transform to compute the survival probability. In the following, we consider this method, which exhibits an exponential convergence, as a benchmark. Fang et al. [9] price credit default swaps (CDS) using the COS method, which is based on a cosine series expansion, and deal with the problem of calibrating the Lévy models to market quotes of CDS prices. Finally, Cariboni and Schoutens [6] assume that the underlying process is of pure jump type, and calculate the survival probability by relating it to the price of a binary down-and-out barrier option as in Feng and Linetski's paper [10], considering both partial integro-differential equation and Monte Carlo approaches. For further details, see also the references cited by Cariboni and Schoutens [6] and Fang et al. [9].

The key point of the structural approach is to compute the survival probability. In this article, we show how this can be done using the WH factorization. The latter can be dealt with from two different perspectives, analytical and probabilistic. The first approach is based on the solution of integral equations, whilst the second is based on the Spitzer identity [25].

The first approach, that we call the Z-WH pricing method, is based on a Z-transform which converts the usual backward procedure into a set of independent integral equations of WH type. The procedure can be described as a randomization of the contract expiry according to a geometric distribution with a complex parameter q, see [11, 12][1]. The solution of the WH integral equations resulting from the randomization and depending on q, requires to compute the factorization into a product of two functions, one analytic in an upper complex half plane, the other in a lower.

The second approach, that we call the Z-WH-S method, is related to the Spitzer identity [25]. This identity factorizes the characteristic function of a random walk with independent and identically distributed increments as the product of the characteristic functions of the minimum and of the maximum of the random walk itself, and all random variables are stopped at a geometric random time. The Spitzer identity gives a probabilistic interpretation of these functions in terms of the characteristic function of the extrema of a geometrically stopped random walk. For a recent discussion, see Bingham [4] and the references therein. We consider as an application the pricing of defaultable bonds following the structural approach explained above. The default probability is thus computed applying the factorization algorithm to the Spitzer identity and the Z-transform inversion.

Finally, in this paper we show how the factorization and the inversion can be performed numerically with high speed and accuracy for a general Lévy process. In particular, the Z-transform inversion is performed as suggested by Abate and Whitt [1], and, in addition, we implement it by using the Euler acceleration. The WH factorization is instead based on an idea originally proposed by Henery [14]. Numerical results are presented in order to validate the pricing algorithm. Moreover, a sensitivity analysis of the probability of default with respect to the parameters of the considered Lévy processes is performed. The problem of computing the term structure of the default probability and of the credit spread is also addressed.

The structure of the paper is the following. In Sect. 2, we introduce the problem of computing the default probability and the price of a defaultable zero coupon bond, and we describe the necessary recursive formula. In Sect. 3, we describe how to exploit the FFT to solve efficiently the factorization problem. Finally, in Sect. 4 we validate our procedures with numerical results, taking into consideration both the accuracy and the computational cost, and we analyze the sensitivity of the probability of default and of the credit spread term structures with respect to the model parameters.

2 Default Probability and Credit Derivatives

We model default adopting a structural approach. The firm value S evolves according to a stochastic process and default will occur as soon as the value of the firm falls below a level L before or at maturity T. The boundary L can be related to bond covenants or to an optimality condition based on the value of equity claims or

---

[1]The interpretation of transforms as probabilities is also frequently called the method of collective marks and is usually attributed to van Dantzig [8]. See also Resnick [23, p. 564].

to the minimal firm value required to operate the company. Rather than assuming a standard geometric Brownian motion process as in Merton [21] and Black and Cox [5], here we model the firm value via an exponential Lévy process. This fact allows us to generate a term structure of credit spreads that does not tend to zero as the time horizon shortens. Instead, this cannot happen if we model the firm value according to a diffusion process. In addition, we assume that the default is monitored only at discrete dates, such as bond coupon payment or balance sheet reporting dates. For the aim of simplicity, we also assume that the N monitoring dates are equally spaced, so that $\Delta = T/N$ is the time interval between two successive monitoring dates. In the following, we denote with r the risk-free interest rate and we assume that the underlying asset S pays a continuous dividend q.

Given a standard filtered probability space, for $0 \leq t \leq T$ the firm value process $S(t)$ is represented as

$$S(t) = S(0)e^{X(t)}, \tag{1}$$

where $X(t)$ is a Lévy process, i.e., a stochastically continuous process with independent and stationary increments, uniquely identified by its characteristic function

$$\Psi(\xi; \Delta) = \mathbb{E}_0 \left[e^{i\xi X(\Delta)} \right]. \tag{2}$$

2.1 Default Probability, Defaultable Coupon Bond, and Credit Spread

If L is the default barrier, the default occurs as soon as $S(t) \leq L$, t being a monitoring date. The quantity

$$z = -\log \frac{S(t)}{L} \tag{3}$$

is commonly known as distance to default. We can also define the default time as the first hitting time of the level L:

$$\tau = \min_{j=0,\ldots,N} \{ j\Delta : S(j\Delta) \leq L \} \tag{4}$$

The probability of default $\mathbb{P}(\tau \leq j\Delta)$ is related to the distribution of the minimum value of the underlying asset. Indeed if we define

$$m_i := \min_{j=0,\ldots,i} X(j\Delta), \quad i = 0, \cdots, N, \tag{5}$$

we have

$$\mathbb{P}(\tau \leq j\Delta) = \mathbb{P}(m_j \leq \log L). \tag{6}$$

The probability of default $p = \mathbb{P}(m_N \leq \log L)$, as well as its complement, the survival probability $1 - p$, are the key ingredients to price credit derivatives, like, for example, defaultable bonds and CDS. A defaultable zero-coupon bond written on a risky asset S is a bond which at maturity T pays a unit notional if the asset price stays above the default threshold L, or pays the recovery fraction $R < 1$ of the notional otherwise (R could also be equal to 0)[2]. The standard convention among academics and practitioners assumes that the recovery rate is a constant parameter, even if there is some empirical evidence of a negative correlation between default rates and recovery rate, see, for example, Altman et al. [2]. We will follow the standard industry practice, treating R as a constant parameter. Thus, once the default probability is computed and the recovery rate is assigned, the price $P_d(T)$ of the defaultable zero-coupon with maturity T is

$$P_d(T) = e^{-rT}(1 - p + Rp). \tag{7}$$

Given that the price of the risk-free zero-coupon is simply $P(T) = e^{-rT}$, the credit spread, i.e., the difference of the yield to maturity of the two zero-coupon bonds, defaultable and non-defaultable, with the same maturity, can be computed as

$$s(T) = -\frac{1}{T}\log\frac{P_d(T)}{P(T)} = -\frac{1}{T}\log\big(1 - p(1-R)\big). \tag{8}$$

2.2 Recursive Valuation of the Default Probability

In order to compute the default probability at time T when N equally spaced monitoring dates are considered, we can define the function $v(x, j)$ through the recursion

$$v(x, j) = \int_0^{+\infty} f(z - x; \Delta) v(z, j - 1) dz, \quad j = 1, \ldots, N, \tag{9}$$

$$v(x, 0) = \mathbf{1}_{x>0},$$

where $\mathbf{1}_{x>0}$ is the indicator function which is equal to 1 if $x > 0$, 0 otherwise, and $f(\cdot; \Delta)$ is the transition probability density function of the log-return of the underlying asset. It holds that

$$\mathbb{P}(m_N \leq \log L) = 1 - v(x_0 - \log L, N). \tag{10}$$

See Fusai et al. [12] for details.

---

[2]This is the standard framework of the so-called fractional recovery of face value. Another possibility, not considered here, is the fractional recovery of market value at default.

3 Fast Pricing Methods

The recursion in Sect. 2.2 can be used to compute the probability of default. However, faster and more accurate pricing techniques can be considered. More precisely, in Sect. 3.1, we present the Z-transform approach, which transforms the recursive procedure presented in Sect. 2.2 into the problem of solving independent integral equations. Since these integral equations are of WH type, in Sect. 3.2 we introduce a fast solution method based on the WH factorization. Finally, in Sect. 3.3 we present a different, but related, approach for computing the probability of default based on the Spitzer identity. This approach seems to be the fastest and the most accurate. This fact is numerically discussed in Sect. 4.

In the following, we denote the Fourier transform of a function g with

$$\widehat{g}(\xi) = \mathscr{F}_{x \to \xi}[g(x)](\xi) = \int_{-\infty}^{+\infty} g(x) e^{i\xi x} dx \tag{11}$$

and the inverse transform with

$$g(x) = \mathscr{F}_{\xi \to x}^{-1}[\widehat{g}(\xi)](x) = \frac{1}{2\pi} \int_{-\infty}^{+\infty} \widehat{g}(\xi) e^{-ix\xi} d\xi. \tag{12}$$

We recall that the characteristic function of the Lévy process, which is given in (2) and is assumed to be known analytically, satisfies

$$\Psi(\xi; \Delta) = \mathscr{F}_{x \to \xi}[f(x; \Delta)](\xi; \Delta). \tag{13}$$

Notice the exception of using Ψ in place of $\widehat{f}$. Furthermore, we define the projection operators on the positive ($\mathscr{P}_x^+$) and negative ($\mathscr{P}_x^-$) real axis[3]

$$\mathscr{P}_x^+ g(x) = \mathbf{1}_{x>0} g(x) \quad \text{and} \quad \mathscr{P}_x^- g(x) = \mathbf{1}_{x<0} g(x). \tag{14}$$

Finally, the index $+$ ($-$) denotes a function analytic on an upper (lower) complex half plane including the real axis. For $\xi \in \mathscr{D}$, $\mathscr{D}$ being a suitable overlapping strip of the two half planes, we set

$$\widehat{g_+}(\xi) = \mathscr{F}_{x \to \xi}\left[\mathscr{P}_x^+ g(x)\right](\xi) := \int_0^{+\infty} g(x) e^{i\xi x} dx \tag{15}$$

$$\widehat{g_-}(\xi) = \mathscr{F}_{x \to \xi}\left[\mathscr{P}_x^- g(x)\right](\xi) := \int_{-\infty}^0 g(x) e^{i\xi x} dx. \tag{16}$$

---

[3]For numerical purposes, it is convenient to use the symmetric Heaviside step function $H(x)$ in place of the indicator function $\mathbf{1}_{x>0}$ and $1 - H(x)$ in place of $\mathbf{1}_{x<0}$, the only difference being for $x = 0$, as $H(0) = 1/2 = 1 - H(0)$.

3.1 The Z-Transform Approach

Let us consider the recursive evaluation given in (9) of the function $v(x, j)$ and let us define the (unilateral) Z-transform $V(x, q)$ of $v(x, j)$ as

$$V(x, q) := \sum_{j=0}^{\infty} q^j v(x, j), \quad q \in \mathbb{C}. \tag{17}$$

The Z-transform can be considered a discrete-time relative of the Laplace transform; the reason for its name might be that usually $z = 1/q$ is employed in its definition. If we apply the Z-transform to the recursive equation (9), it can be shown that the Z-transform $V(x, q)$ satisfies the integral equation

$$V(x, q) = q \int_0^{+\infty} f(z - x; \Delta) V(z, q) \mathrm{d}z + \phi(x), \tag{18}$$

with the "forcing" function $\phi(x) := v(x, 0) = \mathbf{1}_{x>0}$. The inverse Z-transform is given by an integral on a closed path around the origin; choosing a circle of radius $\rho < 1$, we have

$$v(x, N) = \frac{1}{2\pi\rho^N} \int_0^{2\pi} V(x, \rho e^{iu}) e^{-iNu} \mathrm{d}u, \tag{19}$$

quantity that can be approximated by

$$v_h(x, N) = \frac{1}{2N\rho^N} \left[V(x, \rho) + 2 \sum_{j=1}^{N-1} (-1)^j \mathrm{Re}\, V(x, \rho e^{ij\pi/N}) + (-1)^N V(x, -\rho) \right]. \tag{20}$$

Setting $\rho = 10^{-\gamma/N}$ leads to an accuracy of $10^{-2\gamma}$ [12]. In our numerical tests, we used $\gamma = 6$.

Therefore, in order to obtain $v_h(x, N)$, one must solve $N + 1$ independent integral equations given by (18) with $N + 1$ different values of the parameter q. Moreover, Fusai et al. [12] proposed to apply the Euler summation or acceleration [22], which is a convergence-acceleration technique well suited to evaluate alternating series. The idea of the Euler summation is to approximate $v_h(x, N)$ by the binomial average (also called the Euler transform)

$$v_h(x, N) \approx \frac{1}{2^{m_E} N \rho^N} \sum_{j=0}^{m_E} \binom{m_E}{j} b_{n_E+j}, \tag{21}$$

of the partial sums

$$b_k = \sum_{j=0}^{k} (-1)^j a_j \mathrm{Re}\, V\left(x, \rho e^{ij\pi/N}\right), \tag{22}$$

ranging from $k = n_E$ to $k = n_E + m_E$, where $a_0 = 1/2$, $a_i = 1$, $i = 1, \ldots,$ and m_E and n_E are suitably chosen such that $m_E + n_E < N$. Thus, the number of integral equations to be solved is $\min(N, m_E + n_E) + 1$. In our numerical experiments, we set $m_E = 20$ and $n_E = 12$.

3.2   The Z-WH Approach

In the previous section, we have shown how the recursive pricing problem presented in Sect. 2.2 can be transformed into the solution of integral equations of WH type, (18). In this section, we introduce an algorithm by Henery [14] well suited to solve this type of integral equations.

The main steps for solving the WH integral equations (18) are:

1. Assign q, the forcing function $\phi(x)$ and the characteristic function $\Psi(\xi; \Delta)$, and define

$$L(\xi, q) := \frac{1}{1 - q\Psi(\xi; \Delta)}, \quad \widehat{\phi_+}(\xi) := \mathscr{F}_{x \to \xi} \left[\mathscr{P}_x^+ \phi(x) \right](\xi). \quad (23)$$

2. Factorize the function $L(\xi, q)$ into a product of two functions (analytical in an upper or lower complex half plane),

$$L(\xi, q) = L_+(\xi, q)L_-(\xi, q), \quad (24)$$

and thus

$$\log L(\xi, q) = \log L_+(\xi, q) + \log L_-(\xi, q). \quad (25)$$

3. Define $C(\xi, q) := L_-(\xi, q)\widehat{\phi_+}(\xi)$ and decompose it into components analytical in the appropriate complex half plane,

$$C(\xi, q) = C_+(\xi, q) + C_-(\xi, q). \quad (26)$$

4. The solution is now given by the inverse Fourier transform

$$V(x, q) = \mathscr{F}_{\xi \to x}^{-1}[C_+(\xi, q)L_+(\xi, q)](x, q). \quad (27)$$

The conditions under which the factorization or decomposition gives proper results are discussed by Krein [15]. In fact, the major difficulty in the analytic solution of the WH equation lies in the factorization, which is known analytically only for a few types of functions. The most important condition for the above scheme is that $q\Psi(\xi; \Delta)$ must not be close to 1 anywhere; otherwise, the function $L(\xi, q)$ to be factorized diverges. Provided this condition is fulfilled, one benefit of using numerical methods is that one is no longer restricted to functions for which analytic factorizations are possible. The second advantage of the proposed method is that the basic building block of this numerical solution of the WH equation is the Fourier transform, which can be implemented conveniently via the FFT.

3.2.1 Factorization

As shown above, the main step of the proposed solution algorithm is the computation of the WH factorization. In fact, in order to implement the solution of the transformed equation, we need to factorize the function $L(\xi,q)$, as in (24), into a product of two functions which are analytic in the overlap of the two half planes. This factorization can be done through the Hilbert transform $\mathscr{H}_\xi$ of $\log L(\xi,q)$ [14],

$$L_+(\xi,q) = \exp\left[\frac{1}{2i}\mathscr{H}_\xi \log L(\xi,q)\right]$$

$$:= \exp\left[\frac{1}{2\pi i}\,\text{p.v.}\int_{-\infty}^{+\infty}\frac{\log L(\xi',q)}{\xi'-\xi}d\xi'\right], \quad \text{Im}\,\xi' < \text{Im}\,\xi,\; \xi' \in \mathscr{D}, \quad (28)$$

where p.v. denotes the principal value, i.e., the value of a complex function along one chosen branch, in order to make the function single valued. It can be shown that, since $\text{Im}(\xi-\xi') > 0$,

$$\frac{1}{2\pi}\int_{-\infty}^{+\infty}\frac{\log L(\xi',q)}{i(\xi'-\xi)}d\xi' = \int_{-\infty}^{+\infty}\mathbf{1}_{x>0}\left(\frac{1}{2\pi}\int_{-\infty}^{+\infty}\log L(\xi',q)e^{-ix\xi'}d\xi'\right)e^{i\xi x}dx.$$

$$(29)$$

Therefore, we have

$$\log L_+(\xi,q) = \mathscr{F}_{x\to\xi}\left[\mathscr{P}_x^+\,\mathscr{F}_{\xi\to x}^{-1}\log L(\xi,q)\right](\xi,q). \quad (30)$$

This expression suggests the use of the FFT as a numerical tool for computing $L_+(\xi,q)$: indeed, given the function $L(\xi,q)$, the factorization of $\log L(\xi,q)$ can be performed through the sequence: inverse Fourier transform, projection on the positive real axis, and Fourier transform. In practice, we have to use twice the FFT algorithm. For details, see Henery [14].

3.2.2 The Algorithm and Its Computational Cost

Summarizing, the solution of the WH equation (18) can be computed as in (27),

$$V(x,q) = \mathscr{F}_{\xi\to x}^{-1}[C_+(\xi,q)L_+(\xi,q)](x,q), \quad (31)$$

where, using the factorization formula (30), C_+ and L_+ can be computed as follows:

1. $L(\xi,q) := 1/(1 - q\Psi(\xi;\Delta))$;
2. $L_+(\xi,q) = \exp\left\{\mathscr{F}_{x\to\xi}[\mathscr{P}_x^+\mathscr{F}_{\xi\to x}^{-1}\log L(\xi,q)](\xi,q)\right\}$;
3. $L_-(\xi,q) = \exp\left\{\mathscr{F}_{x\to\xi}[\mathscr{P}_x^-\mathscr{F}_{\xi\to x}^{-1}\log L(\xi,q)](\xi,q)\right\} = L(\xi,q)/L_+(\xi,q)$;
4. $\widehat{\phi_+}(\xi) := \mathscr{F}_{x\to\xi}[\mathscr{P}_x^+\phi(x)](\xi)$;
5. $C_+(\xi,q) = \mathscr{F}_{x\to\xi}[\mathscr{P}_x^+\mathscr{F}_{\xi\to x}^{-1}[L_-(\xi,q)\widehat{\phi_+}(\xi)]](\xi,q)$.

The computational cost of this Z-WH pricing procedure consists of six Fourier transforms for each integral equation, i.e., $6(\min(N, n_E + m_E) + 1) m \log m$ operations. However, we can easily decrease the computational cost applying the Z-transform inversion directly in the Fourier space, i.e., considering (20) and (31), if we define
$$\widehat{V}(\xi, q) := C_+(\xi, q) L_+(\xi, q), \tag{32}$$
it holds that

$$v_h(x, N) = \frac{1}{2N\rho^N} \mathrm{Re} \, \mathscr{F}^{-1}_{\xi \to x} \left[\widehat{V}(\xi, \rho) + 2 \sum_{j=1}^{N-1} (-1)^j \widehat{V}(\xi, \rho e^{ij\pi/N}) \right.$$

$$\left. + (-1)^N \widehat{V}(\xi, -\rho) \right] (x, N), \tag{33}$$

or, using the Euler summation (21),

$$v_h(x, N) \approx \frac{1}{2^{m_E} N \rho^N} \mathrm{Re} \, \mathscr{F}^{-1}_{\xi \to x} \left[\sum_{j=0}^{m_E} \binom{m_E}{j} b_{n_E + j}(\xi) \right], \tag{34}$$

where

$$b_k(\xi) = \sum_{j=0}^{k} (-1)^j a_j \, \widehat{V}\left(\xi, \rho e^{ij\pi/N}\right). \tag{35}$$

Moreover, the computation of $\widehat{\phi}_+(\xi) = \mathscr{F}_{x \to \xi}[\mathscr{P}_x^+ \phi(x)](\xi)$ can be performed only once. Thus, to compute the default probability $4(\min(N, n_E + m_E) + 1) + 2$ FFTs are necessary.

3.3   The Z-WH-S Approach

In this section, we discuss how to compute the probability of default exploiting the Spitzer identity. The Spitzer identity [25] is strictly related to the solution of WH integral equations. Indeed, in the Z-WH approach integral equations are solved computing a decomposition into two functions, one analytic in an upper complex half plane, the other in a lower complex half plane; the two half planes overlap in a strip that includes the real axis. The Spitzer identity consists in providing a probabilistic interpretation of these two functions in terms of the characteristic function of the minimum of a geometrically stopped random walk. More precisely, the Spitzer identity states that

$$\sum_{j=0}^{+\infty} q^j \mathbb{E}_0 \left[e^{i\xi m_j} \right] = L_+(0, q) L_-(\xi, q), \tag{36}$$

where L_+ and L_- have been defined previously in Sect. 3.2. Thus, once the WH factorization is computed as above, we can invert the Z-transform as in (21) exploiting the Euler acceleration, obtaining an approximation of the characteristic function of the discrete minimum m_N. Finally, applying an inverse Fourier transform, we obtain the probability density function of the minimum $p_{m_N}(x; \Delta)$, i.e., if $N > n_E + m_E$,

$$p_{m_N}(x; \Delta) \approx \frac{1}{2^{m_E} N \rho^N} \operatorname{Re} \mathscr{F}_{\xi \to x}^{-1} \left[\sum_{j=0}^{m_E} \binom{m_E}{j} b_{n_E+j}(\xi) \right], \qquad (37)$$

where

$$b_k(\xi) = \sum_{j=0}^{k} (-1)^j a_j L_+ \left(0, \rho e^{ij\pi/N} \right) L_- \left(\xi, \rho e^{ij\pi/N} \right). \qquad (38)$$

Summarizing, the Z-WH-S algorithm is the following.

1. Compute the WH factorization as suggested above:

   (a) $L(\xi, q) := 1/(1 - q\Psi(\xi; \Delta))$;
   (b) $L_-(\xi, q) = \exp\left\{ \mathscr{F}_{x \to \xi} [\mathscr{P}_x^- \mathscr{F}_{\xi \to x}^{-1} \log L(\xi, q)](\xi, q) \right\}$;
   (c) $L_+(0, q) = L(0, q)/L_-(0, q)$.

2. Compute the inverse of the Z-transform exploiting the Euler acceleration, obtaining the characteristic function of the minimum m_N.
3. Compute the inverse Fourier transform of the characteristic function of the minimum to obtain its distribution $p_{m_N}(\cdot; \Delta)$.
4. Compute the cumulative density function $l \to \mathbb{P}(m_N \leq l)$ and thus the default probability.

In this case, we have to compute only $2(\min(N, n_E + m_E) + 1) + 1$ Fourier transforms.

4 Numerical Experiments

In this section, we present numerical results in order to validate the pricing algorithm. More precisely, in Sect. 4.1 we price a defaultable bond with the Z-WH and the Z-WH-S methods, for different maturities, comparing both the accuracy and the computational costs. As a benchmark, we consider the method of Feng and Linetsky [10]. Moreover, an analysis on the sensitivity of the probability of default and the credit spread with respect to the model's parameters is considered in Sect. 4.2. In our numerical experiments, we set the risk-free rate $r = 0.05$ and we assume that the underlying asset pays a dividend equal to $q = 0.02$. The asset price at time 0 is $S(0) = 1$, the default barrier is $L = 0.3$, and the recovery is $R = 0.5$. All the numerical experiments have been performed with Matlab R2009b running under

Table 1 Characteristic exponents of some parametric Lévy processes

Model	Characteristic exponent
G	$-\frac{1}{2}\sigma^2\omega^2$
NIG	$-\delta\left(\sqrt{\alpha^2-(\beta+i\omega)^2}-\sqrt{\alpha^2-\beta^2}\right)$
CGMY	$C\Gamma(-Y)\left((M-i\omega)^Y-M^Y+(G+i\omega)^Y-G^Y\right)$
DE	$-\frac{1}{2}\sigma^2\omega^2+\lambda\left(\frac{(1-p)\eta_2}{\eta_2+i\omega}+\frac{p\eta_1}{\eta_1+i\omega}-1\right)$
JD	$-\frac{1}{2}\sigma^2\omega^2+\lambda\left(e^{i\omega\alpha-\frac{1}{2}\omega^2\delta^2}-1\right)$

Table 2 Defaultable zero-coupon bonds with weekly monitoring: price and CPUtime (in seconds)

		Z-WH			Z-WH-S			HILB		
T	m	p (%)	Price	CPUt	p (%)	Price	CPUt	p (%)	price	CPUt
1	2^{12}	0.9335	0.946789	0.17	0.9329	0.946792	0.13	0.9330	0.946791	0.29
1	2^{14}	0.9330	0.946791	0.66	0.9330	0.946791	0.48	0.9330	0.946791	0.58
2	2^{12}	4.5656	0.884181	0.17	4.5650	0.884184	0.14	4.5652	0.884183	0.35
2	2^{14}	4.5652	0.884183	0.68	4.5652	0.884183	0.46	4.5652	0.884183	0.88
5	2^{12}	21.5713	0.694801	0.18	21.5702	0.694806	0.13	21.5703	0.694805	0.58
5	2^{14}	21.5703	0.694805	0.69	21.5703	0.694805	0.43	21.5703	0.694805	1.99
10	2^{12}	42.9653	0.476321	0.17	42.9663	0.476228	0.14	42.9670	0.476226	0.73
10	2^{14}	42.9624	0.476240	0.70	42.9670	0.476226	0.44	42.9670	0.476226	3.45

Windows 7 on a personal computer equipped with an Intel Dual-Core 2.70 GHz processor and 4 GB of RAM. The characteristic exponents of the considered Lévy processes are reported in Table 1.

4.1 Numerical Validation

In Table 2, we report the default probabilities p and the defaultable zero-coupon bond prices considering different maturities T and 52 monitoring dates a year. We assume that the underlying asset evolves as an exponential Normal-Inverse Gaussian (NIG) process with the same parameters as those used by Feng and Linetsky [10], i.e., $\alpha = 5$, $\beta = -1$, $\gamma = 0.75$. Finally, we denote with m the number of grid points used to compute the FFT. Our results are tested considering as a benchmark the method of Feng and Linetsky [10] based on the Hilbert transform (HILB).

As expected, the HILB method rapidly reaches a six decimal digits accuracy due to its exponential convergence [10]. However our methods show a similar accuracy, and have a computational cost independent of the number of monitoring dates, as shown in Table 2. Thus, the Z-WH and Z-WH-S algorithms appear to perform better than the HILB method when more than 2 years with weekly monitoring (i.e., 104 monitoring dates) are considered. Finally, the Z-WH-S method appears to outperform the Z-WH algorithm in both accuracy (especially with a small number of grid points) and speed. We recall that the Z-WH-S method requires approximately $2\min(N, m_E + n_E)\,m\log m$ operations, while the Z-WH $4\min(N, m_E + n_E)\,m\log m$.

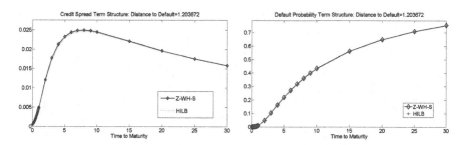

Fig. 1 Credit spread (left) and default probability (right) term structure

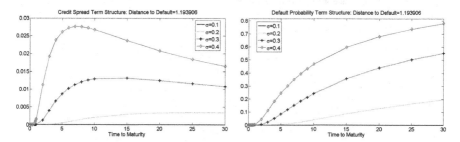

Fig. 2 Term structure of the credit spread (left) and of the default probability (right) when the log-firm value evolves according to a Gaussian process

4.2 Credit Spread Term Structure

In Fig. 1, we show the default probability and the credit spread term structures. We consider a weekly monitoring and 30 different maturities from 2 weeks to 30 years, comparing the HILB method with the Z-WH-S method with $m = 2^{14}$ grid points. The two methods provide the same term structure; however, the CPU time necessary to obtain the term structures is 9.35 s with the Z-WH-S algorithm and 47.82 s with the HILB method.

In Fig. 2, we show the default probability and the credit spread term structures, assuming that the underlying asset evolves according to a Gaussian process with different values of the volatility σ. The term structures are computed with the Z-WH-S method with $m = 2^{14}$ grid points. In this figure, we can see how both the default probability and the credit spread shift up as the parameter σ increases, as we should expect.

In Fig. 3, we report similar plots considering the NIG process and four different sets of parameters:

1. $\alpha = 5, \quad \beta = -1, \quad \gamma = 0.75;$
2. $\alpha = 2, \quad \beta = -1, \quad \gamma = 0.75;$
3. $\alpha = 5, \quad \beta = -0.5, \gamma = 0.75;$
4. $\alpha = 5, \quad \beta = -1, \quad \gamma = 1.$

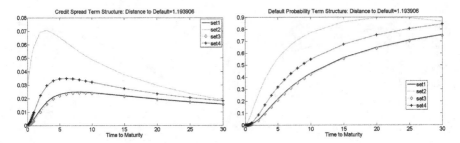

Fig. 3 Term structure of the credit spread (left) and of the default probability (right) when the log-firm value evolves according to an NIG process

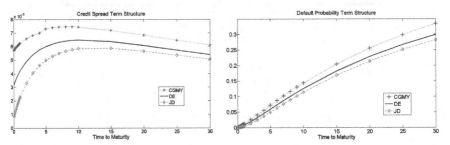

Fig. 4 Term structure of the credit spread (left) and of the default probability (right) when the log-firm value evolves according to a CGMY, DE, or JD process

We recall that for the NIG process the variance is given by $\alpha^2 \delta (\alpha^2 - \beta^2)^{-3/2}$. Thus, the four sets of parameters have variance equal to 0.1599, 0.5774, 0.1523, and 0.2126, respectively. In Fig. 3, we see a behavior similar to the one in Fig. 2: increasing the variance, both term structures shift up.

Finally in Figs. 4 and 5 we compare the term structures for different Lévy processes, assuming again a weekly monitoring. More precisely, in Fig. 4, in order to make the term structures comparable across models, the parameters (see Table 1) have been chosen assuming that the CGMY model, as estimated by Schoutens [24], is the true one:

$$C = 0.0244, \ G = 0.0765, \ M = 7.5515, \ Y = 1.2945.$$

Therefore, we calibrate the other models by minimizing the square integrated difference between the real part of the characteristic functions of the CGMY and the other models. Thus the calibrated parameters for the Merton jump diffusion (JD) model are $\sigma = 0.126349$, $\alpha = -0.390078$, $\lambda = 0.174814$ and $\delta = 0.338796$, while the calibrated parameters for the Kou double exponential (DE) model are $\sigma = 0.120381$, $\lambda = 0.330966$, $p = 0.20761$, $\eta_1 = 9.65997$ and $\eta_2 = 3.13868$. The risk-free rate is 3.67% per year, and the dividend yield is set equal to zero.

We recall that considering non-Gaussian processes allows us to generate term structures of credit spreads that do not tend to zero as the time horizon shortens: this can be easily seen in Fig. 4.

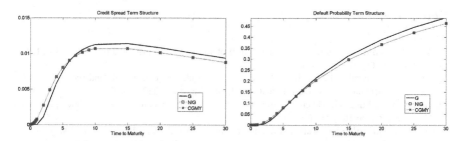

Fig. 5 Term structure of the credit spread (left) and of the default probability (right) when the log-firm value evolves according to a CGMY, NIG, or Gaussian (G) process

In Fig. 5, we consider the parameters used by Černý and Kyriakou [7], where the models are calibrated to achieve a volatility 0.3, and, in addition, for the non-Gaussian distributions, skewness -0.5 and kurtosis 3.7; more precisely, for the Gaussian process (G) we set $\sigma = 0.3$, for the NIG model $\alpha = 12.34$, $\beta = -5.8831$, $\gamma = 0.7543$, and finally for the CGMY model $C = 0.6509$, $G = 5.853$, $M = 18.27$, $Y = 0.8$. The risk-free rate is 4% per year, and the dividend yield is set equal to zero.

Figure 5 shows that, once the different Lévy processes are calibrated to share the same first four moments, the term structures of credit spreads and default probabilities are very similar. Therefore, the model risk related to the choice of different exponential Lévy processes is quite limited. We also notice a difference with respect to the Gaussian case. For example, from Fig. 5 we observe a higher (lower) probability of default and a higher (lower) credit spread for small (large) maturities. This behavior is still related to the presence of jumps that generate a different behavior as the time horizon shortens with respect to the Gaussian case.

5 Conclusions

In this article, adopting a structural approach and assuming that the firm value evolves according to an exponential Lévy process, we have proposed two algorithms for computing the default probability based on the Wiener–Hopf factorization: the Z-WH method, in which a set of integral equations has to be solved, and the Z-WH-S ones, which is based on the Spitzer identity. We have shown that the two methods are accurate and fast, and that they are convenient when a large number of monitoring dates is considered, since the computational cost is independent of this number thanks to the Euler acceleration which improves the Z-transform inversion. Moreover, we have provided numerical experiments to validate the algorithms. Finally, we have shown how to compute the probability of default, the price of defaultable zero-coupon bonds, and the corresponding term structure of credit spreads, performing also some comparative analysis across different Lévy processes.

References

1. Abate, J., Whitt, W.: Numerical inversion of probability generating functions. Oper. Res. Lett. **12**, 245–251 (1992)
2. Altman, E.I., Brady, B., Resti, A., Sironi, A.: The link between default and recovery rates: theory, empirical evidence and implications. J. Bus. **78**, 2203–2227 (2005)
3. Atkinson, C., Fusai, G.: Discrete extrema of Brownian motion and pricing of exotic options. J. Comput. Finance **10**, 1–43 (2007)
4. Bingham, N.H.: Random walk and fluctuation theory. Handb. Stat. **19**, 171–213 (2001)
5. Black, F., Cox, J.C.: Valuing corporate securities: some effects of bond indenture provisions. J. Finance **31**, 351–367 (1976)
6. Cariboni, J., Schoutens, W.: Pricing credit default swaps under Lévy models. J. Comput. Finance **10**, 1–21 (2007)
7. Černý, A., Kyriakou, I.: An improved convolution algorithm for discretely sampled Asian options. Quant. Finance **11**, 381–389 (2010)
8. Dantzig, D.V.: Sur la methode des fonctions generatrices. Colloques Internationaux du CNRS **13**, 29–45 (1948)
9. Fang, F., Jönsson, H., Oosterlee, C.W., Schoutens, W.: Fast valuation and calibration of credit default swaps under Lévy dynamics. J. Comput. Finance 14, 57–86 (2010)
10. Feng, L., Linetsky, V.: Pricing discretely monitored barrier options and defaultable bonds in Lévy process models: a Hilbert transform approach. Math. Finance **18**, 337–384 (2008)
11. Fusai, G., Marazzina, D., Marena, M.: Pricing discretely monitored Asian options by maturity randomization. SIAM J. Financ. Math. **2**, 383–403 (2011)
12. Fusai, G., Marazzina, D., Marena, M., Ng, M.: Z-transform and preconditioning techniques for option pricing. Available online 15 Apr 2011. DOI 10.180/14697688.2010.538074 Quant. Finance, in press (2012). Available online 15 Apr 2011. DOI 10.1080/14697688.2010.538074
13. Green, R., Fusai, G., Abrahams, I.D.: The Wiener-Hopf technique and discretely monitored path-dependent option pricing. Math. Finance **20**, 259–288 (2010)
14. Henery, R.J.: Solution of Wiener-Hopf integral equations using the fast Fourier transform. J. Inst. Math. Appl. **13**, 89–96 (1974)
15. Krein, M.G.: Integral equations on a half-line with kernel depending on the difference of the arguments. T. Am. Math. Soc. **22**, 163–288 (1963)
16. Kudryavtsev, O., Levendorskiĭ, S.: Fast and accurate pricing of barrier options under Lévy processes. Finance Stoch. **13**, 531–562 (2009)
17. Kuznetsov, A.: Wiener-Hopf factorization and distribution of extrema for a family of Lévy processes. Ann. Appl. Probab. **20**, 1801–1830 (2010)
18. Kuznetsov, A., Kyprianou, A.E., Pardo, J.C.: Meromorphic Lévy processes and their fluctuation identities. Ann. Appl. Probab., in press (2012)
19. Kuznetsov, A., Kyprianou, A.E., Pardo, J.C., van Schaik, K.: A Wiener-Hopf Monte-Carlo simulation technique for Lévy processes. Ann. Appl. Probab. 21, 2171–2190 (2011)
20. Kyprianou, A.E.: Introductory Lectures on Fluctuations of Lévy Processes with Applications. Springer-Verlag, Berlin (2006)
21. Merton, R.C.: Theory of rational option pricing. Bell J. Econ. Manage. Sci. **4**, 141–183 (1973)
22. O'Cinneide, C.A.: Euler summation for Fourier series and Laplace transform inversion. Stoch. Models **13**, 315–337 (1997)
23. Resnick, S.: Adventures in Stochastic Processes. Birkhäuser Boston (2002)
24. Schoutens, W.: Lévy Processes in Finance. Wiley Chichester (2003)
25. Spitzer, F.A.: A combinatorial lemma and its application to probability theory. T. Am. Math. Soc. **82**, 327–343 (1956)

The Evaluation of Gas Swing Contracts with Regime Switching

Carl Chiarella, Les Clewlow, and Boda Kang

Abstract A typical gas swing contract is an agreement between a supplier and a purchaser for the delivery of variable daily quantities of gas, between specified minimum and maximum daily limits, over a certain period at a specified set of contract prices. The main constraint of such an agreement that makes them difficult to value are that there is a minimum volume of gas (termed take-or-pay or minimum bill) for which the buyer will be charged at the end of the period (or penalty date), regardless of the actual quantity of gas taken. We propose a framework for pricing such swing contracts for an underlying gas forward price curve that follows a regime-switching process in order to better capture the volatility behavior in such markets. With the help of a recombining pentanomial tree, we are able to efficiently evaluate the prices of the swing contracts and find optimal daily decisions in different regimes. We also show how the change of regime will affect the decisions.

1 Introduction

In the natural gas markets, there is an increasing focus on swing contracts and gas storage assets as sources of hidden, untapped flexibility. The best practice accountancy and management of flexible gas assets require a thorough understanding of the underlying gas market fundamentals, and the range of supporting mathematical techniques for the assets' evaluation and optimization. An inadequate understanding of these issues could result in the suboptimal performance of flexible assets, in both

C. Chiarella (✉) • B. Kang
Finance Discipline, UTS Business School, University of Technology, Sydney,
PO Box 123, Broadway, NSW 2007, Australia
e-mail: carl.chiarella@uts.edu.au; boda.kang@uts.edu.au

L. Clewlow
Lacima Group, Sydney, Australia
e-mail: les@lacimagroup.com

M. Cummins et al. (eds.), *Topics in Numerical Methods for Finance*, Springer Proceedings
in Mathematics & Statistics 19, DOI 10.1007/978-1-4614-3433-7_9,
© Springer Science+Business Media New York 2012

financial and physical terms. In this paper, we mainly concentrate on the evaluation of the gas swing contracts.

Clewlow et al. [11, 12] discuss the risk analysis and the properties of the optimal exercise strategies with the help of a trinomial tree method. Barrera-Esteve et al. [2] develop a stochastic programming algorithm to evaluate swing options with penalty. Bardou et al. [1] use the so-called optimal quantization method to price swing options with the spot price following a mean reverting process.

Most recently, [14] implement a pentanomial lattice approach to evaluate swing options in gasoline markets under the assumption that the spot price follows a regime-switching Geometric Brownian Motion where the volatility can switch between different values based on the state of a hidden Markov chain.

Breslin et al. [5] introduced the definition and explained many basic features of a typical gas swing contract, which is an agreement between a supplier and a purchaser for the delivery of variable daily quantities of gas—between specified minimum and maximum daily limits—over a certain number of years at a specified set of contract prices. While swing contracts have been used for many years to manage the inherent uncertainty of gas supply and demand, it is only in recent years with deregulation of the energy markets that there has been any interest in understanding and valuing the optionality contained in these contracts. In the model of [5], the volatility is a deterministic function of both the current time and the time-to-maturity; however, there is a great deal of evidence indicating that the volatility is stochastic in gas markets and we argue that a regime-switching model is better able to capture such random features. The main contribution of this paper is to evaluate a single year gas sales agreement (GSA) or swing contract introduced in [5], but with a regime-switching forward price curve.

The remainder of the paper is structured as follows. In Sect. 2, we propose a one-factor regime-switching model for the gas forward price curve and we build a recombining pentanomial tree to approximate the gas spot price process derived from the forward price curve model. We introduce the basic features and the detailed evaluation procedures of a single year GSA in Sect. 3. In Sect. 4, we provide several numerical examples to demonstrate the properties of both the decision surfaces and value surfaces of these contracts. We draw some conclusions in Sect. 5.

2   Regime-Switching Forward Price Curve and a Tree

Forward contracts are widely traded on many exchanges with prices easily observed—often the nearest maturity forward price is used as a proxy for the spot price with longer dated contracts used to imply the convenience yield. Clewlow and Strickland [9] work in this class of models, simultaneously modeling the evolution of the entire forward curve conditional on the initially observed forward curve and so setup a unified approach to the pricing and risk management of a portfolio of energy-derivative positions. In this paper, we follow this approach to model the volatility functions of the forward curve directly.

2.1 Forward Price Curve with Regime-Switching Volatility

Deterministic volatility functions cannot capture the complicated movements of the forward curves. Hence, we propose a stochastic volatility model under which we price a single year GSA. Volatility of the forward curve is stochastic due to a hidden Markov Chain that causes it to switch between "high volatility load" and "low volatility load" states. Chiarella et al. [8] have found that a regime-switching model captures quite well the stochastic nature of the volatility function in the gas market and they implement an MCMC approach to estimate the parameters of the model.

In this paper, we consider a one-factor regime-switching forward curve model:[1]

$$\frac{\mathrm{d}F(t,T)}{F(t,T)} = \langle \sigma, X_t \rangle c(t) \cdot \mathrm{e}^{-\alpha(T-t)} \mathrm{d}W_t, \tag{1}$$

where

- $F(t,T)$ is the price of the gas forward at time t with a maturity at time T.
- W_t is a standard Brownian Motion.
- The time-varying term $c(t) = c + \sum_{j=1}^{M} (d_j(1 + \sin(f_j + 2\pi jt)))$ captures the seasonal effect.
- X_t is a finite state Markov chain with state space $E = \{e_1, e_2, \ldots, e_N\}$ where e_i is a vector of length N and equal to 1 at the ith position and 0 elsewhere, that is,

$$e_i = (0, \ldots, 0, 1, 0, \ldots, 0)' \in \mathbb{R}^N,$$

where $'$ indicates the transpose operator.
- $P = (p_{ij})_{N \times N}$ is the transition probability matrix of the Markov Chain X_t. For all $i = 1, \ldots, N, j = 1, \ldots, N$. The quantity p_{ij} is the conditional probability that the Markov Chain X_t transits from state e_i at current time t to state e_j at the next time $t + \Delta t$, that is,

$$p_{ij} = \Pr(X_{t+\Delta t} = e_j | X_t = e_i).$$

- $\sigma = (\sigma_1, \sigma_2, \cdots, \sigma_N)$ are the different values of the volatilities which evolve following the rule of the Markov Chain X_t.
- $\langle \cdot, \cdot \rangle$ denotes the scalar product in $\mathbb{R}^N$, if $\sigma = (\sigma_1, \cdots, \sigma_N)$ then

$$\langle \sigma, X_t \rangle = \sum_{i=1}^{N} \sigma_i \mathbf{1}_{(X_t = e_i)},$$

---

[1]Chiarella et al. [8] suggest a two-factor regime-switching forward curve model. However, for simplicity and for the purpose of demonstrating the implementation of the pentanomial tree approach we use a one-factor model here. We also argue that a multifactor model is necessary for hedging and risk management purposes a one-factor model is a good approximation for pricing. Also, it is not hard to generalize the approach in the paper to handle two or more factor models but the computational effort will of course increase.

where the indicator function is

$$\mathbf{1}_{(X_t = e_i)} = \begin{cases} 1, & \text{if } X_t = e_i; \\ 0, & \text{otherwise.} \end{cases}$$

This scalar product allows the spot volatility of the forward price curve to switch among different values σ_i randomly depending on the state of the Markov Chain X_t.

We also know that for $F(t,T)$ satisfying Eq. (1) the spot price $S(t) = F(t,t)$ is given by (see, e.g., [5])

$$S(t) = F(0,t) \cdot \exp\left(\int_0^t \langle \sigma, X_s \rangle c(s) \cdot e^{-\alpha(t-s)} dW_s - \frac{1}{2} \Lambda_t^2 \right), \tag{2}$$

where $\Lambda_t^2 = \int_0^t (\langle \sigma, X_s \rangle c(s) \cdot e^{-\alpha(t-s)})^2 ds$.

2.2 Pentanomial Tree Construction

The spot price dynamics in Eq. (2) is rather complicated since it involves the path dependence of the history of the hidden Markov chain which makes it hard to construct a recombining discrete grid to approximate the continuous spot price process. The single year GSA that we are trying to evaluate has several features and also can be early exercised multiple times during the life of the contract. We are aware that the Least-Squares Monte Carlo method (LSMC) has been used to evaluate both the gas storage contract, see, e.g., [4, 7] and the gas swing contract without penalty, see, e.g., [13]. However, the penalty at the end of the gas year introduces a discontinuity in the first derivative of the value surface not only in the spot price dimension but also in the volume taken dimension and this aspect the LSMC does not handle well. Furthermore, the complexity of multi-year swing contracts with features such as make-up and carry-forward provisions cannot be handled easily by LSMC because of the additional discontinuities.

We have found that lattice approaches are widely used because of their computational simplicity and flexibility. Bollen [3] constructed a pentanomial lattice to approximate a regime-switching Geometric Brownian Motion. Wahab and Lee [14] extended the pentanomial lattice to a multinomial tree and studied the price of swing options under regime-switching dynamics. Those researchers study a different version of the contract which has multiple early exercise opportunities without penalty. However, the penalty usually will be applied at the end of a gas year for a single year contract and the above-mentioned discontinuity introduced by the penalty makes the contract difficult to evaluate. Wahab and Lee [14] did discuss pricing a swing option under a regime-switching model but there are two main differences between their paper and ours:

1. Wahab and Lee [14] proposed a Geometric Brownian Motion process for the gasoline price where the volatility can switch between different regimes according to a Markov chain. However, the mean reverting process in our paper is more appropriate in capturing the behavior of the gas price and the process of building a pentanomial tree for a regime-switching mean reverting process is different to that for GBM.

2. A very important feature of the gas swing contract is that there will be a penalty at the end of the gas year if the volume taken in the year did not meet the minimal bill; hence, both the penalty and the volume taken make the contracts more complicated to evaluate. However, all the contracts in the numerical examples in [14] neither have a penalty nor take the volume taken into consideration which makes the features of the contract essentially different from what we have in our paper.

In this paper, in order to construct a discrete lattice that approximates the spot price process $S(t)$, we let $Y_t = \int_0^t \langle \sigma, X_s \rangle c(s) \cdot e^{-\alpha(t-s)} dW_s$, so that

$$dY_t = -\alpha Y_t dt + \langle \sigma, X_t \rangle c(t) dW_t, \tag{3}$$

and we build a discrete lattice to approximate Y_t first. Then at each time step, we add an adjustment term to the nodes on the lattice for Y_t so that the lattice obtained for the spot price is consistent with the observed market forward price curve (as detailed below).

2.2.1 Nodes

We assume that there are only two regimes ($N = 2$) for the volatility. Instead of σ_1 and σ_2, we use σ_L when $X_t = L$ for the low volatility regime and σ_H when $X_t = H$ for the high volatility regime. In the one-stage pentanomial tree in Fig. 1, each regime is represented by a trinomial tree with one branch being shared by both regimes. In order to minimize the number of nodes in the tree, the nodes from both regimes are merged by setting the step sizes of both regimes at a 1 : 2 ratio which is the only ratio to make the tree recombine when we have two regimes[2]. Figure 2 demonstrates the recombing nature of the tree.

The time values represented in the tree are equally spaced and have the form $t_j = j\Delta t$ where j is a nonnegative integer and Δt is the time step, usually one day in our context. The values of Y at time t_j are equally spaced and have the form $Y_{j,k} = k\Delta Y$ where ΔY is the space step and k determines the level of the variable in the tree. Any node in the tree can therefore be referenced by a pair of integers, that is, the node at the jth time step and kth level we refer to as node (j,k). From stability and convergence considerations, a reasonable choice for the relationship

---

[2]This ratio needs to be adjusted accordingly if we have $N > 2$ regimes, see [14] for more details.

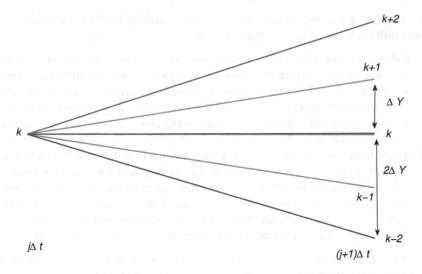

Fig. 1 One step of a pentanomial tree. The outer two branches together with the middle branch represent the regime with high volatility and the inner two branches together with the middle branch represent the regime with low volatility

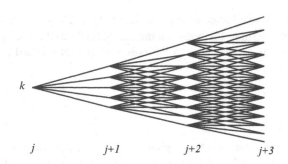

Fig. 2 The recombining nature of a pentanomial tree

between the space step ΔY and the time step Δt suggested by Wahab and Lee [14] is given by

$$\Delta Y = \begin{cases} \sigma_L \sqrt{3\Delta t}, & \sigma_L \geq \sigma_H/2, \\ \frac{\sigma_H}{2} \sqrt{3\Delta t}, & \sigma_L < \sigma_H/2. \end{cases}$$

The trinomial branching process and the associated probabilities are chosen to be consistent with the drift and volatility of the process. The three nodes that can be reached by the branches emanating from node (j,k) are $(j+1,l-1)$, $(j+1,l)$, and $(j+1,l+1)$ for the low volatility regime and $(j+1,l-2)$, $(j+1,l)$, and $(j+1,l+2)$ for the high volatility regime. Here l is chosen so that the value of Y reached

Fig. 3 The alternative branching processes for mean reverting processes. The level where the tree switches from one branch to another depends on the attenuation parameter α and the time step Δt

by the middle branch is as close as possible to the expected value of Y at time t_{j+1}.[3] From the Euler discretization of Eq. (3), the expected value of Y at time t_{j+1} conditional on $Y = Y_{j,k}$ is $Y_{j,k} - \alpha Y_{j,k}\Delta t$.

2.2.2 Transition Probabilities

For either regime $x = L$ or H, let $p_{u,j,k}^{x}, p_{m,j,k}^{x}$, and $p_{d,j,k}^{x}$ define the probabilities associated with the upper, middle, and lower branches emanating from node (j,k), respectively. These probabilities can be calculated as follows. When the volatility is in the low regime, $\sigma = \sigma_L$, looking at the inner trinomial tree we need to match

$$E[\Delta Y] = -\alpha Y_{j,k}\Delta t, \text{ and } E[\Delta Y^2] = \sigma_L^2 c(t_j)\Delta t + E[\Delta Y]^2.$$

Therefore equating the first and second moments of ΔY in the tree with the above values, we obtain

$$p_{u,j,k}^{L}((l+1)-k) + p_{m,j,k}^{L}(l-k) + p_{d,j,k}^{L}((l-1)-k) = -\alpha Y_{j,k}\frac{\Delta t}{\Delta Y}, \qquad (4)$$

$$p_{u,j,k}^{L}((l+1)-k)^2 + p_{m,j,k}^{L}(l-k)^2 + p_{d,j,k}^{L}((l-1)-k)^2$$
$$= (\sigma_L^2 c(t_j)\Delta t + (-\alpha Y_{j,k}\Delta t)^2)/\Delta Y^2. \qquad (5)$$

---

[3]The determination of l depends on the different alternative branching processes for the mean reverting processes. For instance, in Fig. 3, from left to right, l is equal to $k-2$, $k-1$, k, $k+1$, and $k+2$, respectively.

Solving Eqs. (4) and (5) together with conditions that $p^L_{u,j,k} + p^L_{m,j,k} + p^L_{d,j,k} = 1$, we obtain

$$p^L_{u,j,k} = \frac{1}{2}\left[\frac{\sigma^2_L c(t_j)\Delta t + \alpha^2 Y^2_{j,k}\Delta t^2}{\Delta Y^2} + (l-k)^2 - \frac{\alpha Y_{j,k}\Delta t}{\Delta Y}(1-2(l-k)) - (l-k)\right],$$

$$p^L_{d,j,k} = \frac{1}{2}\left[\frac{\sigma^2_L c(t_j)\Delta t + \alpha^2 Y^2_{j,k}\Delta t^2}{\Delta Y^2} + (l-k)^2 + \frac{\alpha Y_{j,k}\Delta t}{\Delta Y}(1+2(l-k)) + (l-k)\right],$$

$$p^L_{m,j,k} = 1 - p^L_{u,j,k} - p^L_{d,j,k}.$$

When the volatility is in high regime, $\sigma = \sigma_H$, looking at the outer trinomial tree and applying a similar procedure, we find that

$$p^H_{u,j,k} = \frac{1}{8}\left[\frac{\sigma^2_H c(t_j)\Delta t + \alpha^2 Y^2_{j,k}\Delta t^2}{\Delta Y^2} + (l-k)^2 - \frac{\alpha Y_{j,k}\Delta t}{\Delta Y}(2-2(l-k)) - 2(l-k)\right],$$

$$p^H_{d,j,k} = \frac{1}{8}\left[\frac{\sigma^2_H c(t_j)\Delta t + \alpha^2 Y^2_{j,k}\Delta t^2}{\Delta Y^2} + (l-k)^2 + \frac{\alpha Y_{j,k}\Delta t}{\Delta Y}(2+2(l-k)) + 2(l-k)\right],$$

$$p^H_{m,j,k} = 1 - p^H_{u,j,k} - p^H_{d,j,k}.$$

2.2.3 State Prices for Both Regimes

Following a similar approach to that in Chapter 7 of [10], we displace the nodes in the above simplified tree by adding the drifts a_i which are consistent with the observed forward prices.

In fact, since we have two regimes, for $x = L, H$ we define state (or Arrow–Debreu) prices $Q^x_{j,k}$ as the present value of a security that pays off \$1 if $Y = k\Delta Y$ and $X_{j\Delta t} = x$ at time $j\Delta t$ and zero otherwise. The $Q^x_{j,k}$ are in fact the state prices that accumulate according to

$$Q^L_{0,0} = 1, \ Q^H_{0,0} = 0 \text{ for the lower volatility regime,}$$

$$Q^L_{0,0} = 0, \ Q^H_{0,0} = 1 \text{ for the higher volatility regime,}$$

$$Q^L_{j+1,k} = \sum_{k'}\left(Q^L_{j,k'} p_{L,L} + Q^H_{j,k'} p_{H,L}\right) p^L_{k',k} B(j\Delta t, (j+1)\Delta t),$$

$$Q^H_{j+1,k} = \sum_{k'}\left(Q^L_{j,k'} p_{L,H} + Q^H_{j,k'} p_{H,H}\right) p^H_{k',k} B(j\Delta t, (j+1)\Delta t),$$

where $p_{x,x'}$ is the probability the Markov Chain transits from the state x to the state x' and $p^L_{k',k}$, and $p^H_{k',k}$ are the probabilities the spot price transits from k' to k but arriving at low and high volatility regimes, respectively. $B(j\Delta t, (j+1)\Delta t)$ denotes the price at time $j\Delta t$ of the pure discount bond maturing at time $(j+1)\Delta t$.

We see that Arrow–Debreu securities are the building blocks of all securities; in particular, when we have j time steps in the tree, the price today, $C(0)$, of any European claim with payoff function $C(S)$ at time step j in the tree is given by

$$C(0) = \sum_k \left(Q_{j,k}^{L} + Q_{j,k}^{H} \right) C(S_{j,k}), \tag{6}$$

where $S_{j,k}$ is the time t_j spot price at level k and the summation takes place across all of the nodes k at time j.

In order to use the state prices to match the forward curve, we use the special case of Eq. (6) that values the initial forward curve, namely

$$B(0, j\Delta t) F(0, j\Delta t) = \sum_k \left(Q_{j,k}^{L} + Q_{j,k}^{H} \right) S_{j,k}. \tag{7}$$

By the definition of a_j we have $S_{j,k} = e^{Y_{j,k} + a_j}$, then the term a_j needed to ensure that the tree correctly returns the observed futures curve is given by

$$a_j = \ln \left(\frac{B(0, j\Delta t) F(0, j\Delta t)}{\sum_k (Q_{j,k}^{L} + Q_{j,k}^{H}) e^{Y_{j,k}}} \right). \tag{8}$$

In fact, inserting $S_{j,k} = e^{Y_{j,k} + a_j}$ into Eq. (7), we have

$$B(0, j\Delta t) F(0, j\Delta t) = \sum_k (Q_{j,k}^{L} + Q_{j,k}^{H}) e^{Y_{j,k} + a_j} = e^{a_j} \sum_k (Q_{j,k}^{L} + Q_{j,k}^{H}) e^{Y_{j,k}}.$$

Hence, we have

$$e^{a_j} = \frac{B(0, j\Delta t) F(0, j\Delta t)}{\sum_k (Q_{j,k}^{L} + Q_{j,k}^{H}) e^{Y_{j,k}}},$$

then Eq. (8) follows immediately.

The upper panel of Fig. 4 demonstrates an example of a pentanomail tree which has been constructed to be consistent with the seasonal gas forward prices shown in the lower panel of Fig. 4.

3 Single Year Gas Sales Agreement

3.1 The Contract

GSA is an agreement between a supplier and a purchaser for the delivery of variable daily quantities of gas, between specified minimum and maximum daily limits, over a certain period at a specified set of contract prices. The main features of these contracts that make them difficult to value and risk manage are the constraints on the quantity of gas which can be taken. The main constraint is that at the end of the

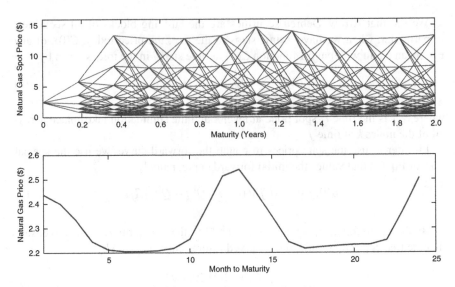

Fig. 4 Spot price tree fitted (*upper panel*) to seasonal forward curve (*lower panel*)

period, there is a minimum volume of gas (termed take-or-pay or minimum bill) for which the buyer will be charged at the end of the year (or penalty date), regardless of the actual quantity of gas taken. Typically, there is also a maximum annual quantity which can be taken. The minimum bill or take-or-pay level is usually defined as a percentage of the notional annual quantity which is called the annual contract quantity (ACQ).

With the help of the pentanomial tree that we have constructed, we are able to evaluate the price of the above swing contract. The value of the contract at maturity (the final purchase date) can be computed first. The final decision at maturity is simple because the penalty amount is known with certainty. Then we step back through the pentanomial tree computing the discounted expectations of the contract value at each node for both low and high volatility regimes and computing the optimal purchase decision at the purchase dates for both regimes as well. The optimal purchase decision at each node and for each value of the remaining volume and for each regime can be computed by searching over the range of possible purchase volumes for the volume which maximizes the sum of the discounted expectation averaged by the transition probabilities of the hidden Markov Chain on different regimes and the value of the current purchase.

3.2 The Evaluation

In the following, we assume that the economy is in regime $x = L, H$ at any particular time depending on the evolution of the hidden Markov chain.

Let $V_t^*(S,Q,x)$ and $q_t^*(S,Q,x), t = 0,1,\ldots,T$ be the time t optimal value and decision function of a Take-or-Pay contract when the spot price is $S_t = S$, the volume taken is $Q_t = Q$, and the system is in regime $x = L,H$. Let MB denote the minimum bill and K be the contract price.

In such contract, the buyers will face, at the end of the year, a penalty of the following form

$$\min\{Q_T - MB, 0\} \cdot K, \tag{9}$$

which is the shortfall not meeting the minimal bill multiplied by the strike price.

Hence, the buyers of the swing contract should take decisions so that their total expected discounted payoffs are maximized. Generally speaking, the buyer decides on each possible trading day whether to exercise one swing right or not, and the amount $q_t^*(S,Q,x)$ taken upon exercise.

In the following, we will give an analysis on the optimal decisions on the last day of the contract. Then the dynamic programming principle will be implemented to work out both the optimal decisions and the optimal values of the swing contract at each day.

Because of the penalty in Eq. (9), we can work out that the optimal decisions $(q_T^*(S,Q,x))$ and the optimal value functions $(V_T^*(S,Q,x))$ at the maturity of the contract are as follows

$$q_T^*(S,Q,x) = \begin{cases} 1, & S > K; \\ \min(\max(MB-Q,0),1), & S \leq K. \end{cases}$$

$$V_T^*(S,Q,x) = (S-K)q_T^*(S,Q,x) - K\max(0,MB-(Q+q_T^*(S,Q,x))).$$

In fact, at the last day of the contract, when the option is in the money $(S>K)$, the buyer should take as much as allowed which is 1 in this case. However when the option is out of money, the buyer does not need to take any if the volume taken exceeds the minimal bill. However, it may be optimal for the buyer to take the amount necessary to avoid the penalty if the volume taken is lower than the minimal bill, in which case the buyer will face the penalty given by Eq. (9). This analysis applies to both regimes.

We step back through the pentanomial tree computing the discounted expectations of the contract value at each node for both low and high volatility regimes and computing the optimal purchase decision at the purchase dates for both regimes as well. The optimal decisions and optimal value functions in the swing period before maturity will be calculated according to Bellman's Optimality Principle or dynamic programming. For $t = T-1, \cdots, 0$, working backward in time, we have:

$$V_t^*(S,Q,x) = \max_{q\in[0,1]} \left\{ q(S-K) + \mathrm{e}^{-r\Delta t} \sum_{x'=L}^{H} p_{xx'}^X \mathbb{E}_S^x[V_{t+1}^*(S_{t+1},Q+q,x')] \right\};$$

$$q_t^*(S,Q,x) = \text{argmax}_q \left\{ q(S-K) + e^{-r\Delta t} \sum_{x'=L}^{H} p_{xx'}^X \mathbb{E}_S^x [V_{t+1}^*(S_{t+1}, Q+q, x')] \right\}.$$

In fact, the optimal purchase decision at each node and for each value of the remaining volume and for each regime can be computed by searching over the range of possible purchase volumes for the volume which maximizes the sum of the discounted expectation averaged by the transition probabilities of the hidden Markov Chain on different regimes and the value of the current purchase.

We also need the following boundary conditions:

$$V_t^*(S, Q_{max}, x) = 0, \ q_t^*(S, Q_{max}, x) = 0,$$

which because the value function will be zero and there is no gas to use if the volume taken reaches the maximal quantity.

The nodes and transition probabilities of the pentanomial tree constructed in the previous section can be used to calculate the above conditional expectation $\mathbb{E}[\cdot|\cdot]$.

4 Numerical Examples

In this section, we provide a numerical example to demonstrate how we evaluate one-year contracts and how we calculate the optimal decisions on the amount of daily gas consumption.

In the following, we evaluate a one-year GSA according to the following parameter settings:

- Volatilities: $\sigma_L = 0.5, \sigma_H = 1.0$
- Mean reversion rate: $\alpha = 5$
- Interest rate: $r = 0$
- The forward price curve is flat: $F(0,t) = 100, \forall t > 0$
- Contract price: $K = 100$
- Daily take limit: $q_{min} = 0$ and $q_{max} = 1$
- Maturity time: $T = 365$
- Minimal Bill: $MB = 365 \times 80\% = 292$
- The maximum annual quantity: $Q_{max} = 365$
- Transition matrix of the hidden Markov Chain: $P1 = \begin{bmatrix} 0.99 & 0.01 \\ 0.01 & 0.99 \end{bmatrix}$
- Transition matrix of the hidden Markov Chain: $P2 = \begin{bmatrix} 0.85 & 0.15 \\ 0.71 & 0.29 \end{bmatrix}$

Following the detailed procedures described in Sect. 2, we build a pentanomial lattice part of which is shown in Fig. 5. It is consistent with the flat forward price curve.

The surfaces in Figs. 6–8 illustrate the key features of a typical decision surface. We denote the critical price as the spot price above which it is optimal to take the maximum daily quantity. In this "Day 0" (pricing date) plot note that at low

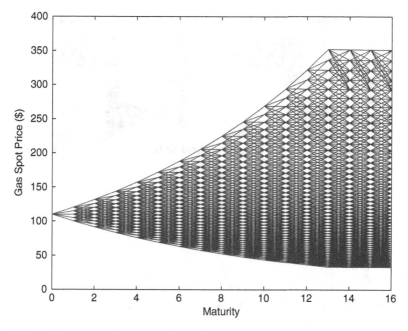

Fig. 5 Part of the pentanomial tree based on a flat forward price curve

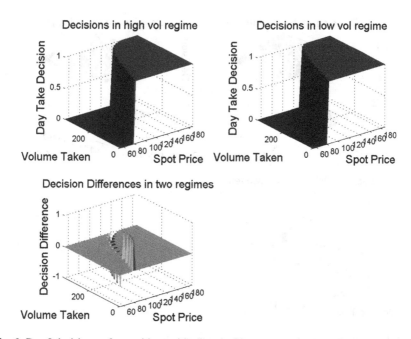

Fig. 6 Day 0 decision surfaces with transition matrix $P1$

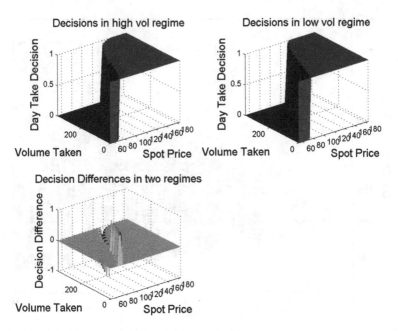

Fig. 7 Day 73 decision surfaces with transition matrix $P1$

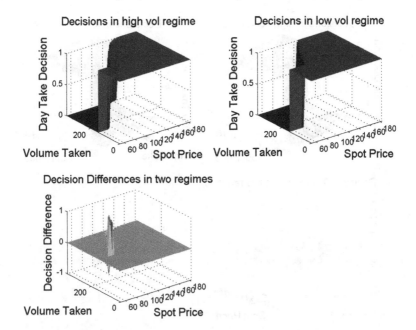

Fig. 8 Day 219 decision surfaces with transition matrix $P1$

"volume taken" values, the critical spot price may be less than the contract strike price, even though this results in an immediate loss. This non-intuitive result is due to the expectation that future spot prices can be even lower and cause greater forced losses in order to meet the value MB. So it is better to take a small loss now to reduce the probability of having to take a larger loss later. Below the critical price, the expectation is that the loss incurred by taking gas today is greater than the expected future loss from taking when the spot price is higher. This is affected by the mean reversion rate—if the mean reversion rate is high then the expected number of lower spot prices decreases, if the reversion rate is low then the expected number of lower spot prices increases.

Notice that the "Day 0" surface also gives the optimal decisions for "volume taken" values greater than zero. This would apply if the buyer and seller of the contract were to agree to an effective existing volume of gas having been taken. As the "volume taken" increases the critical price also increases. This is because the maximum annual take is 365 so the holder of the contract will only be able to take gas on at most (365—"volume taken") days. This is reflected in the higher critical price which indicates that the holder must be more selective about which days to take gas. In effect having fewer days on which they can take gas means there will be more days they can choose not to take gas, which will obviously be those days with low relative prices.

For a nonzero MB condition there will be a point in time at which the decisions become constrained and the optimal decision is to take gas irrespective of the spot price. For this example, the MB condition of 80 % first shows at Day 73 (20 % into the year), and the effect is clear at Day 219 (60 % into the year) where the critical price is at zero for all "volume taken" levels below 40 % of the maximum annual take (146). This level of "volume taken" leaves 40 % of the maximum annual take to reach MB in the 40 % of the year remaining.

In Figs. 9–11, we show the corresponding value surfaces for the same 3 days as above figures showing decision surfaces. The plot at Day 0 illustrates a typical unconstrained value surface, where highest value is found for high spot price and low "volume taken" values. This is where the expectation is highest for positive future cashflows—spot prices much greater than the contract price. Conversely, the lowest value occurs for the lowest spot price and lowest "volume taken"—this is where the expectation is highest for negative future cashflows.

As we move forward in time and the system becomes constrained by the Minimum Bill condition (beyond day 73), the impact on the value surface becomes clear—values in the constrained region start to decrease. In these cases, as the "volume taken" volume decreases the values become more negative. This arises because of the MB constraint which means the buyer is forced to pay for the gas even if it is not taken. This can be viewed as a penalty that is imposed when the total gas take for the contract is less than the MB level. The effect of this is clearly evident at day 219.

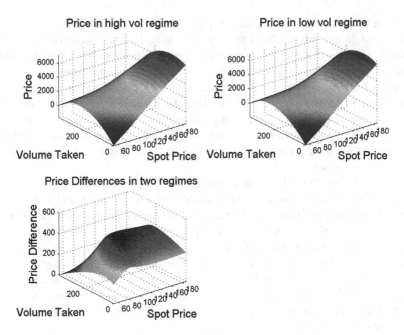

Fig. 9 Day 0 value surfaces with transition matrix $P1$

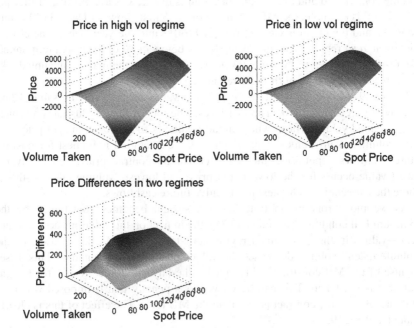

Fig. 10 Day 73 value surfaces with transition matrix $P1$

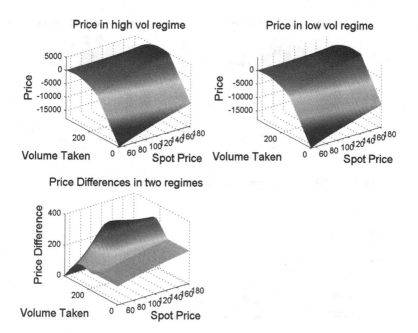

Fig. 11 Day 219 value surfaces with transition matrix $P1$

At the final time step applying to both transition matrices $P1$ and $P2$ in Figs. 12 and 13, if the "volume taken" value is greater than MB the value is zero; but if it is less than MB, then the penalty is proportional to the difference between the MB and the volume of gas taken.

Figures 14–19 demonstrate both the decision and value surfaces when the transition matrix of the hidden Markov chain is $P2$ instead of $P1$. The system will stay in both states sufficiently long when the transition matrix takes the form of $P1$ but the system will mainly stay in state 1 for most of the time and occasionally jump to state 2 when the transition matrix takes the form of $P2$. The differences between two regimes for both value and decisions are much smaller when the transition matrix takes the form $P2$.

5 Conclusions

In this paper, we propose a pentanomial tree framework for pricing one-year swing contracts for an underlying gas forward price curve that follows a regime-switching process. The swing contracts are complicated because the buyer can exercise his or her rights in a daily manner. Hence, in the evaluation we need to keep track of multiple variables on a daily basis. Those complexities, along with the regime-switching uncertainty of the daily price, require efficient numerical procedures to

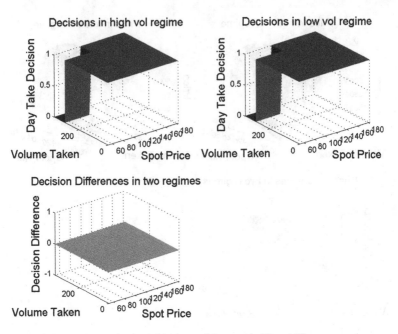

Fig. 12 Day 365 decision surfaces with both transition matrix $P1$ and $P2$

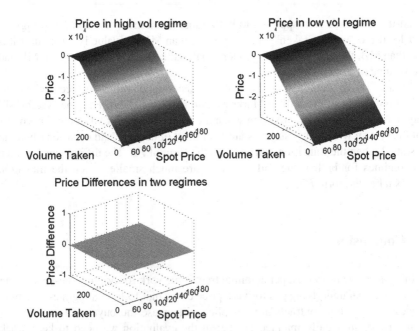

Fig. 13 Day 365 value surfaces with both transition matrix $P1$ and $P2$

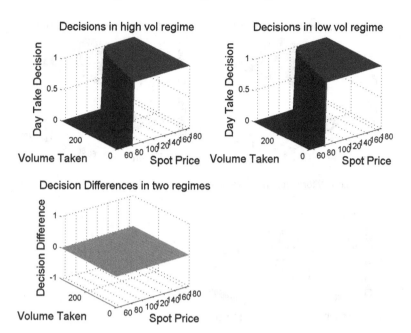

Fig. 14 Day 0 decision surfaces with transition matrix $P2$

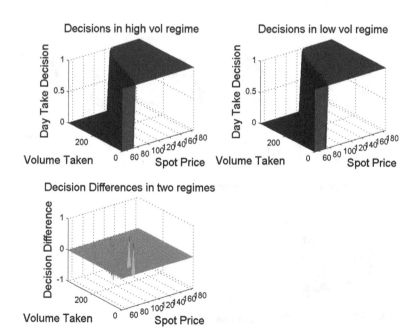

Fig. 15 Day 73 decision surfaces with transition matrix $P2$

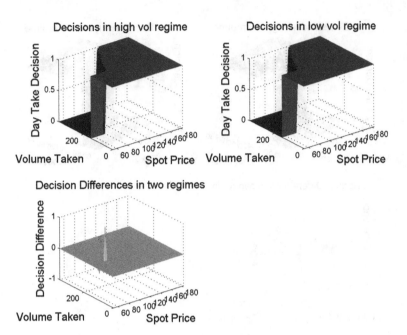

Fig. 16 Day 219 decision surfaces with transition matrix $P2$

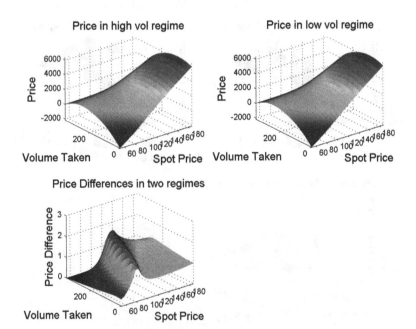

Fig. 17 Day 0 value surfaces with transition matrix $P2$

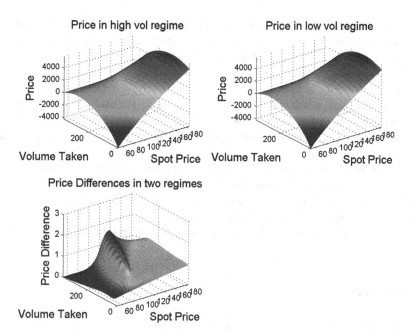

Fig. 18 Day 73 value surfaces with transition matrix $P2$

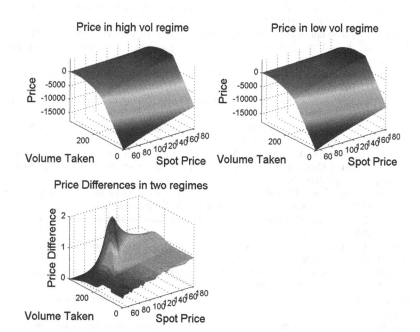

Fig. 19 Day 219 value surfaces with transition matrix $P2$

value these contracts and have been the main contribution of this paper. With the help of a recombining pentanomial tree, we are able to efficiently evaluate the prices of the contracts, and find optimal daily decisions in different regimes.

This example also shows that even the intrinsic strategy, which assumes that the future spot price is known, is complex to solve. However, the intrinsic strategy unlocks only part of the value in a swing contract. The full value can only be realized when taking account of the uncertainty in spot prices, and can only be determined using an efficient and sophisticated numerical solver to evaluate all possible decisions.

Breslin et al. [6] discuss the risks and hedging of swing contracts with the features we have discussed in this paper. Hence, an important task of future research will be to find the risks and the hedging strategies for these contracts when the underlying forward curve follows regime-switching dynamics. The computational tools developed in this paper will play an important role in this research agenda.

References

1. Bardou, O., Bouthemy, S., Pagès, G.: Optimal quantization for the pricing of swing options. Appl. Math. Finance **16**(2), 183–217 (2009)
2. Barrera-Esteve, C., Bergeret, F., Dossal, C., Gobet, E., Meziou, A., Munos, R., Reboul-Salze, D.: Numerical methods for the pricing of swing options: a stochastic control approach. Meth. Comput. Appl. Probab. **8**, 517–540 (2006)
3. Bollen, N.: Valuing options in regime-switching models. J. Derivatives **6**, 38–49 (1998)
4. Boogert, A., Jong, C.D.: Gas storage valuation using a monte carlo method. J. Derivatives **15**(3), 81–98 (2008)
5. Breslin, J., Clewlow, L., Strickland, C., van der Zee, D.: Swing contracts: take it or leave it? Energy Risk February 2008 64–68 (2008a)
6. Breslin, J., Clewlow, L., Strickland, C., van der Zee, D.: Swing contracts part 2: risks and hedging. Energy Risk March 2008 56–60 (2008b)
7. Carmona, R., Ludkovski, M.: Valuation of energy storage: an optimal switching approach. Quant. Finance **10**(4), 359–374 (2010)
8. Chiarella, C., Clewlow, L., Kang, B.: Modelling and estimating the forward price curve in the energy market. Quantitative Finance Research Centre, University of Technology Sydney. Working Paper No. 260 (2009)
9. Clewlow, L., Strickland, C.: Valuing energy options in a one factor model fitted to forward prices. QFRC Research Paper Series 10, University of Technology, Sydney (1999a)
10. Clewlow, L., Strickland, C.: Energy Derivatives-Pricing and Risk Management, Lacima Group, London (2000)
11. Clewlow, L., Strickland, C., Kaminski, V.: Valuation of swing contracts in trees. Energy and Power Risk Management, **6**(4) 33–34, (2001a)
12. Clewlow, L., Strickland, C., Kaminski, V.: Risk analysis of swing contracts. Energy and Power Risk Management, **6**(5) 32–33, (2001b)
13. Kiesel, R., Gernhard, J., Stoll, S.: Valuation of commodity-based swing options. J. Energy Markets **3**(3), 91–112 (2010)
14. Wahab, M., Lee, C.: Pricing swing options with regime switching. Ann. Oper. Res. **185**(1), 139–160 (2009)

A Linear and Nonlinear Review of the Arbitrage-Free Parity Theory for the CDS and Bond Markets

Kitty Moloney and Srinivas Raghavendra

Abstract The arbitrage-free parity theory states that there is equivalence between credit default swap (CDS) spreads and bond market spreads in equilibrium. We show that the testing of this theory through the application of linear Gaussian bivariate modeling will lead to misleading results for CDS and bond spreads, and that linear stochastic modeling is not appropriate for CDS spreads. We propose the nonlinear and nonparametric dynamic tools of cross recurrence plots and cross recurrence plot measures to evaluate the arbitrage-free parity theory. We conclude that convergence is nonmean reverting and varying through time and across countries. This finding refutes the arbitrage-free parity theory. We also conclude that the probability to arbitrage will be affected by country and time-specific factors such as the expectation for country-specific government intervention. We propose that this methodology could be used by policy markets to supervise arbitrage activity and to influence policy making.

1 Introduction

A credit default swap (CDS) contract gives the buyer protection against any losses due to a credit event occurring for the underlying entity, [1]. The premium the buyer pays for this protection is known as the spread. A bond market spread is the premium of the bond yield above the risk-free rate. Both spreads should represent the probability to default of the underlying entity, [2]. The arbitrage-free parity theory states that there will be equivalence between CDS spreads and bond market spreads (also known as credit spreads) in equilibrium. Deviations from equivalence

K. Moloney (✉) • S. Raghavendra
J.E. Cairnes School of Business and Economics, National University
of Ireland Galway, Galway, Ireland
e-mail: kitty.moloney@nuigalway.ie

M. Cummins et al. (eds.), *Topics in Numerical Methods for Finance*, Springer Proceedings   177
in Mathematics & Statistics 19, DOI 10.1007/978-1-4614-3433-7_10,
© Springer Science+Business Media New York 2012

will disappear as traders take advantage of the opportunity for arbitrage profits. The assumption that the arbitrage-free parity theory holds in equilibrium has lead to the use of CDS spreads to estimate an implied probability of default when pricing bond yields and vice versa, [3] . This assumption also supports the strategy of hedging bond positions with CDS contracts and vice versa.

There are a number of economic explanations for the existence of nonequivalence between the CDS and bond spread during a crisis. Fontana and Scheicher [4] highlight that as of March 2010, euro zone sovereign CDS markets represent less than 8% of the credit of the underlying sovereign bond market and therefore the CDS market is much smaller in terms of volume and liquidity. During a crisis, they suggest that there will be a "flight to liquidity," which will cause the inequivalence between the sovereign CDS spread and the sovereign bond spread to be maintained. Traders will be unwilling to enter into the required arbitrage trade (to remove the inequivalence) as they will be reluctant to hold the more illiquid CDS positions. Duffie [5], has suggested that institutional impediments rise during a crisis: for example, search costs for trading counterparties can rise as can the time it takes to raise capital. These rising institutional impediments can inhibit trading activities. Mitchel and Pulvino [6] also suggest that capital moves slowly during a crisis and that this may restrict trading activity and prevent hedge fund managers from taking advantage of arbitrage opportunities.

A number of papers attempt to empirically test this theory. Blanco et al., [7], confirm the equivalence of CDS and bond spreads for corporate entities as an equilibrium condition, using linear stochastic models. They also note strong evidence of a lack of convergence in the short run and in 10% of cases a lack of convergence in the long run. They suggest the short-run deviations are due to the CDS spread leading the bond spread in the price-discovery process. The long-run deviations they suggest are due to imperfections in the specifications of the CDS contract (making them unsuitable for arbitrage trading) or are due to measurement error.

Prior to the financial crisis, the expectation of default of sovereign bonds was extremely low. But as the crisis developed, this expectation increased and we saw higher CDS spreads and bond yields. This led to further academic review of the factors influencing sovereign CDS and bond markets as well as the analysis of equivalence between the two assets. Fontana and Scheicher [4] and Haugh et al., [8] apply linear stochastic models to a number of deductively suggested factors influencing the CDS and bond spreads. They conclude that both markets appear to be influenced by common factors. Delis and Mylonidis, [9] replicate some of the work of Blanco et al., [7] but for the sovereign markets. They conclude that except in times of high risk, there is evidence of CDS spreads Granger causing changes in bond spreads. In these high-risk times, the causality goes both ways, indicating the flight to safety of investors from the higher risk sovereign bonds to German bunds.

The objective of this paper is to update this work, testing the arbitrage-free parity theory for CDS and bond spreads using linear and nonlinear tools. We will be focusing on four euro zone peripheral countries which have recently experienced increased default risk due to the financial crisis, that is, Ireland, Spain, Portugal, and Greece. The time period we analyze is from November 2008 until December 2010. This will allow us to review the effect of the sovereign crisis which gained pace during the first few months of 2010. May 2nd, 2010 saw the first Greek bailout by the EU and the IMF. May 9th, 2010 saw the foundation of the European Financial Stability Facility (EFSF) to support economic stability in the euro zone area. May 14th, 2010 saw the ECB intervene in the euro debt markets (through the Securities Markets Programme), in an attempt to reduce bond yields and CDS spreads. November 21st, 2010 saw the bailout of Ireland by the EU and the IMF.

The structure of paper is as follows, Sect. 2 will outline the theoretical underpinnings of the analysis. Section 3 will describe the data to be analyzed. Section 4 will present the results of the analysis and Sect. 5 will summarize our conclusions.

2 Theoretical Underpinnings

In the first section of the theoretical underpinnings (i.e., Sect. 2.1), we apply a linear stochastic test, that is, the Granger causality test, to see if we can find a linear lagged relationship between sovereign CDS spreads and sovereign bond spreads. We also analyze the distributional characteristics of the CDS and bond spread data, statistically analyzing the descriptive statistics and applying the Jarque–Bera test for normality. Section 2.2 starts with a brief introduction to the theoretical framework of nonlinear time series analysis. In Sect. 2.2–2.4, we apply nonlinear time series tools to test the nonlinear structure of the equivalence between CDS spreads and bond spreads.

2.1 Granger Causality and Descriptive Statistics

The majority of the analysis of equivalence between CDS and bond spreads applies linear dynamic stochastic models, [4, 7–9]. As an illustration and for comparison purposes, we apply the linear Granger causality test, [12]. This is a bivariate linear regression test, applying lagged values of both variables as regressors. For example,

$$y_t = \alpha_0 + \alpha_1 y_{t-1} + \alpha_2 y_{t-2} + \beta_1 x_{t-1} + \beta_2 x_{t-2} + \varepsilon_t \quad (1)$$

A Wald's f test is then applied to test the hypothesis that

$$H_0 = \beta_1 = \beta_2 = 0 \quad (2)$$

Accepting the null implies that the x_t variable does not Granger cause changes in y_t. We choose a default value of 2 lags in all cases. We initially allow the CDS spread to be the dependent variable (i.e., y_t) and allow the bond spread to be the independent variable (i.e., x_t). Acceptance of the null implies that bond spreads do not Granger cause CDS spreads. We then swap the positions of the assets, allowing the bond spread to be the dependent variable (i.e., y_t) and the CDS spreads to be the independent variable (i.e., x_t). In this case, acceptance of the null implies that CDS spreads do not Granger cause bond spreads. We compare our results to those of previous literature. We note that as the data are initially transformed to logged differentials prior to applying the test that we do not need to test for cointegration, [11].

Following this comparison, we will review the distributional characteristics of the CDS and bond spreads in each country to see if the assumption of Gaussian residuals is reasonable. We will also apply statistical t tests and f tests to compare the equality of the estimated means and variances of the two assets in each of the four countries.

The drawback of linear modeling is that all irregular behavior has to be due to some random input, [10]. Also, parameters are assumed to be constant over time. All dependencies in the data must be included in the linear model so that the residuals can be assumed to be iid (i.e., independent and identically distributed), otherwise stochastic modeling is not appropriate, [11]. In general, a Gaussian distribution is assumed, this imposes restrictions on the forecast outcomes of the model (e.g., we assume that approx. 99% of residuals must be within $\pm 3\sigma$ of the mean). We will review these requirements in the results section.

2.2 The BDS Test

As mentioned above, in order to use stochastic modeling, we must assume that the data are iid (i.e., independent and identically distributed). In order to review the previous literature in this field, we want to examine this assumption. The BDS test is a test of independence, but unlike its linear counterparts (e.g., the LB Q test) it tests for nonlinear as well as linear dependencies in the data. With the BDS test, we analyze the trajectory of the data in three dimensions (i.e., a three-dimensional phase plane). If we find that the trajectory remains close over a period of time, this indicates that the data are not independent but in fact is attracted to certain points or cycles (known as attractors). If the trajectory is not close over time, this indicates that there is no dependency structure in the data and that the data are independent. We apply the BDS to the residuals of the autoregressive generalized autoregressive conditional heteroskedastic (ARGARCH) model. This model is commonly used to model financial data as it allows for autocorrelation in the mean and the variance. As illustrated below, the ARGARCH model successfully removes all linear dependencies in the data. Thus, a rejection of the null of the BDS test indicates that there are remaining nonlinear dependencies in the data. If this is the case, stochastic modeling may not be appropriate.

The BDS test [14] applies the correlation integral ($C_{m,n}(\varepsilon)$) to test for dependence in a time series. Firstly, we convert the time series (x_t) into a series of vectors (X_t^m). The embedding theorems tell us that if only a few dominant dimensions remain in the system, we can reconstruct the motion of the system using the phase space of a single variable. The embedding dimension m is generally chosen between a value of 2–6 for the BDS test. The value of m determines the number of scalar points in the vector as follows:

$$X_t^m = (x_t, x_{t+1}, \ldots, x_{t+m-1})$$ (3)

The integral $(C_{m,n}(\varepsilon))$ is then estimated, it measures the spatial correlation for the particular embedded dimension, m. The integral can be interpreted as the probability that (for the dimension m) the vector length $\|X_t^m - X_s^m\|$ is less than or equal to ε, a predetermined distance.

$$C_{m,n}(\varepsilon) = \frac{2}{n(n-1)} \sum_t \sum_{s,s<t} I_{[0,\varepsilon]}(\|X_t^m - X_s^m\|)$$ (4)

Such that I(.) denotes the Heaviside function, which takes either the value of 0 or 1 according to

$$I_{[0,\varepsilon]}(s) = \begin{cases} 1 & \text{if } s \in [0,\varepsilon] \\ 0 & \text{if } s \notin [0,\varepsilon] \end{cases}$$ (5)

and $\|.\|$ denotes the supremum norm, given by

$$\|u\| = \sup_{i=1,\ldots,m} |u_i|$$ (6)

If we calculate a high value for the correlation integral, this suggests that the data are not independent. Brock et al., [14] show that at the limit, the integral should follow a scaling principle. That is for:

$$\lim_{n \to \infty} C_{m,n}(\varepsilon) = C_m(\varepsilon)$$ (7)

If the data are independent then:

$$C_m(\varepsilon) = [C_1(\varepsilon)]^m$$ (8)

From this generalized rule, they develop the standardized BDS test statistic (T) as follows:

$$T = \frac{\sqrt{n}(C_m(\varepsilon) - C_1(\varepsilon)^m)}{s_m(\varepsilon)}$$ (9)

such that n is the number of observations, and $s_m(\varepsilon)$ is a consistent estimator of the asymptotic standard deviation $\sigma_m(\varepsilon)$ of $\sqrt{n}(C_m(\varepsilon) - C_1(\varepsilon)^m$, [14]. Brock [14] shows that the test statistic is normally distributed with $N(0,1)$. Dependence is found if the BDS test statistics is significantly different from the z statistic of the normal distribution.

To apply the test, we need to first remove the linear dependencies in the data. We take each sample of data separately and apply the ARGARCH model, as follows:

$$r_t = \mu + \gamma r_{t-1} + \varepsilon_t$$ (10)

$$\varepsilon_t = \sigma_t z_t$$ (11)

such that

$$z_t \sim N(0,1) \tag{12}$$

$$\sigma_t^2 = \alpha_0 + \alpha_1 \varepsilon_{t-1}^2 + \beta \sigma_{t-1}^2 \tag{13}$$

We note that r_t is the logged differential of the series (the spread) at time t (we transform the data to remove the nonstationary trend), μ, γ, α_0, α_1, β are all parameter values. ε_t is the conditional stochastic term. σ_t is the conditional standard deviation, which is dependent on a constant and ε_{t-1}^2, the lagged residual squared and σ_{t-1}^2 the lagged variance. We assume z_t follows a standard normal distribution.[1] In our previous paper [13], the BDS test was applied to ARGARCH residuals of the iTRaxx European index of corporate CDS contracts. Nonlinear dependence was shown to remain in the residuals, thus questioning the use of stochastic modeling in the analysis of CDS contracts. We again apply this methodology here, to examine for nonlinearities in the sovereign CDS and bond spreads.

2.3 Cross Recurrence Plots

As we wish to examine the equivalence between CDS spreads and bond spreads without making any parametric assumptions, we apply the nonlinear time series tool; cross recurrence plots (CRPs) and cross recurrence plot (CRP) measures. Recurrence plots were proposed by Eckmann et al., [17] to evaluate deterministic characteristics of systems. As mentioned above in Sect. 2.2, when considering a dynamical deterministic system, we note that the trajectory of the system will be attracted to certain points or cycles. During these times, the system will revisit the same area of the phase plane. Thus by visually analyzing the recurrence in the embedded time series, we can evaluate if the system is revisiting certain areas of the plane. Continuing recurrence over time is an indication that the system is following a recurring trajectory. This would indicate the existence of attractors and suggest a deterministic system.

A recurrence plot (RP) is a visual representation of recurrence in the system. As RPs make no assumptions about the model underlying the system itself, they can be used to analyze nonstationary systems without parametric assumptions. For these reasons, RPs are particularly useful in the analysis of financial time series. In this paper, we will be focusing on a development within the RQA literature, that is, the CRP and the CRP statistical measures, [18]. This method compares two time series, to see if there is equivalence between them.

As in Sect. 2.2, in order for this methodology to be applied, we need to convert the series of scalars into a series of vectors [19,20]. The time series, x_n, is embedded,

---

[1]One small additional point is that in order to ensure the test has the nuisance parameter free property, that is, that the test can be applied to the residuals of a model, the residuals of the ARGARCH process are first standardized, and transformed into the logged squared residuals [15,16].

choosing an embedding dimension m and an additional parameter the time delay, τ, [21], as follows:

$$X_i = (x_i, x_{i+\tau}, \ldots, x_{i+(m-1)\tau}), i = 1 \cdots n \qquad (14)$$

We will simultaneously embed another time series, as above to create the vector series Y_n, we can test the closeness of each point of the first trajectory, X_n with each point of the second trajectory, Y_n, as follows:

$$CR_{i,j} = I(\varepsilon_i - \| X_i - Y_j \|), X_i, Y_j \in \Re^m, i.j = 1 \cdots n \qquad (15)$$

where n is the number of considered states X_i and Y_j, ε_i is a threshold distance, $\|.\|$ a norm and $I(.)$ the Heaviside function [22].

In this test, we allow the CDS spread data to be converted to a series of vectors (X_i) and the bond spread data to be converted to a series of vectors (Y_j). This is a similar transformation as in the BDS test, except in this test we have an additional parameter, the time delay, τ. We include the time delay here for correctness; in fact with discrete data it is preferable to allow the time delay to equal 1 [23]. In this case there is no difference between the transformation in (3) and (14).

The embedding dimension m is chosen using the method of false nearest neighbors, and for continuous data, the mutual information method is used to estimate τ, [10]. As mentioned above, for discrete empirical data, it is best to take a value of 1 for τ, so that no data points are skipped, [23]. With regard to the threshold distance ε, it has been noted by Marwan [24], that "a general and systematic study on the recurrence threshold selection remains an open task." Many methods have been suggested, the key is to choose a method which maximizes the signal detection, [25]. One method is to choose the threshold equal to 5% of the maximal phase space diameter, [26]. This method did not fit well with our empirical data, as it led to too high a level of recurrence. We choose a threshold level which keeps the number of recurrences low, relative to the number of points in the system, approximately 1%. This improves the likelihood that we are analyzing recurrence due to deterministic behavior, [27]. The same parameters are applied across all the series to ensure consistency of results.

The application of (15) gives $CR_{i,j}$ an $n \times n$ binary matrix. The CRP is a visual representation of the binary matrix, with a black dot indicating a 1, and a white space indicating a 0. By examining the plot for structure, we can interpret the trajectories of the equivalence through time. Depending on the nature of the underlying process, certain patterns will appear.

CRPs can also be represented as distance plots by allowing the distance between the two series to be represented by a changing color or lightening of the plot from black to white (therefore we do not choose a threshold value). We illustrate below in Fig. 1a (on the left), a CRP for two randomly generated time series; the plot is represented by a series of random black dots. We can see no lines within the plot; this indicates that the data are randomly moving from one position to another. With dynamical systems, we expect to see structure in the plot. As Eckmann et al., [29] highlighted, small scale structures such as diagonal lines parallel to the main diagonal occur within dynamical systems. A diagonal line parallel to the main

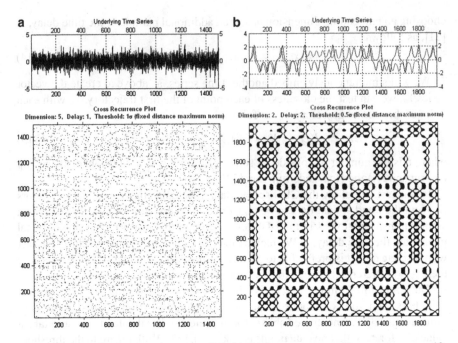

Fig. 1 CRPs. (**a**) Two random data series (L.H.S.) (**b**) Lorenz system with parameters, $\sigma = 10$, $r = 28$, and $b = 8/3$, [30] compared with itself lagged one period forward

diagonal line indicates that a certain trajectory is repeated at different times. For example, the plot of a sine wave will be a series of diagonal lines, as the time series repeats the same trajectory over and over again. In Fig. 1b (on the right), we have a plot of two nonlinear chaotic systems (identical except one is lagged by a time delay of 1), known as the Lorenz system of equations, [10]. This plot shows a checkerboard-like pattern, Eckmann et al., [29] noted that this pattern indicates that the trajectories are moving around attractors. In general, we also note that white bands generally represent two points/or vectors which are far apart. This is an indication of extreme events or nonstationarity in the system [28].

2.4 Convergence and Common Dynamics

CRP measures are the quantification of the patterns in the CRP, through statistical values, [31, 32]. Analysis of these statistical values can indicate many surprising characteristics of a process. For example, they can indicate dynamical convergence and common dynamics between the two time series, [24]. Convergence between two states occurs when their respective phase space trajectories become very close.

For CDS and Bond spreads, this would occur when the values of each are close as follows:

$$d(x - y) \leq \varepsilon; \forall t \tag{16}$$

where d is the distance function and ε is some critical distance, [33]. In general, convergence occurs when there is a recurrence. Convergence between CDS and Bond spreads would imply that the markets are similarly valuing the two spreads. Consistent convergence would be supporting evidence for the arbitrage-free parity theory. We measure for convergence by analyzing the distance between the CDS spread and the bond spread over time. If the two assets converge, we would expect the distance between them to be consistently small.

Common dynamics (or synchronization) occurs when the time evolution of two or more states is similar. This can be examined through the analysis of the phase space trajectories of the two states, [34]. If the system is noisy, it may not be possible to visually recognize the synchronization; thus, the use of CRPs and CRP measures allows us to examine the potential common dynamics between two states.

As Huygens discovered in 1665, two pendulum clocks mounted on a rack of finite rigidity will synchronize over time due to the slight rocking of the rack itself. Synchronicity between states is a characteristic of a complex system [10]. A complex system is a network of heterogeneous components that interact nonlinearly and give rise to emergent behavior; complexity is a general term encompassing chaos, fractals, and other nonlinear theories, [30]. Evidence of nonlinear synchronicity between CDS spreads and Bond spreads would support the view that the financial markets are indeed a complex system. The key to synchronization for Huygens was the influence of the rack on the two pendulums. Synchronization between CDS and bond spreads would imply that similar common (possibly macroeconomic or trading) forces are causing changes in the two asset spreads. A lack of synchronization would imply that different forces are influencing the two asset spreads or that the extent of the influence is different, whereas variability in the level of synchronization implies a varying coupling strength between the two states, [34].

Given two variables x_t and y_t, this can be thought of as

$$\psi(x_t) - \psi(y_t) \leq \pm\omega \tag{17}$$

where ψ is a phase function and ω represents a critical phase shift, [33]. If we find that the difference between the two functions is less than the critical phase shift, we can conclude that there are common dynamics occurring. We measure synchronization by analyzing the nonlinear deterministic structure of the CDS spread and the bond spread over time. If the two assets are synchronized, we would expect the deterministic structure to be almost equal.

Marwan and Kurths [34], test CRP measures by firstly examining synchronization between noisy periodic data sets. They compare the classical cross correlation function with the CRP measures and show that both methods indicate the linear synchronization between the two data sets. Secondly they study the nonlinear synchronization between a linear autoregressive stochastic process and the x-component of the Lorenz system. The two equations have a partially similar structure. The coupling strength k determines the impact of the similar elements in the equations on the overall trajectory of each equation. By increasing the value of the coupling strength k, the CRP measures, recurrence rate (RR) and average diagonal line length (L) also increase. This makes them suitable measures to find the nonlinear relation between two data sets. This nonlinear relationship does not show up using the cross correlation analysis. Thus, CRP measures allow us to assess the extent of linear and nonlinear convergence and synchronization between two data series.

CRP measures are calculated by analyzing the distribution of the diagonal line lengths $P_t(l)$ for each diagonal line parallel to the main diagonal. By doing so, we are focusing on the closeness of the two variables through time.

RR is defined as

$$RR(t) = \frac{1}{n-t} \sum_{l=1}^{n-t} lP_t(l) \tag{18}$$

[34].

RR for cross recurrence analysis measures the probability of occurrence of similar states in both systems. A high density of recurrence points results in a high value for RR, [34]. As we are dealing with discrete data we will be choosing a value for the threshold, ε, to keep RR close to 1%. A rising RR indicates that the degree of convergence is increasing. Thus, we will be analyzing the change in RR over time and comparing RR across countries.

L, is the average diagonal line length and allows us to assess the average duration of the common dynamics between the two states. A high coincidence of both systems increases the lengths of these diagonals, [34].

L is defined as

$$L(t) = \frac{\sum_{l=l_{\min}}^{n-t} lP_t(l)}{\sum_{l=l_{\min}}^{n-t} P_t(l)} \tag{19}$$

[34].

Both RR and L are determined as functions of the distance from the main diagonal, [28]. As with RR, we will be analyzing the change in L over time and across countries to assess the nature of the synchronization (i.e., changing common dynamics) between the two data series.[2]

3 Data

As the objective of the paper is to update and review recent literature on the sovereign credit markets, testing for the arbitrage-free parity theory, we have chosen to analyze four peripheral euro zone countries which recently experienced heightened credit risk. These are Ireland, Spain, Portugal, and Greece. We choose

---

[2]Other measure of CRP can be found at www.recurrence-plot.tk/.

sovereign CDS and bonds of five-year maturities as 85% of the CDS market relates to these maturity lengths, [35]. In order to compare our results, it was necessary to ensure that the number of observations in each data sample is the same and covers the same time period. As the expectation of default for sovereign debt has only recently increased from very low levels, the sovereign CDS markets are relatively new. This implied that we are restricted to the number of observations available in the shortest sovereign CDS market. The shortest data set related to the Greek CDS, which was available daily from 19th November, 2008 through our data source, DataStream. Thus each data sample was taken from 19th November, 2008 until 15th December 2010, a sample size of 541 observations.

As discussed above, the arbitrage-free parity theory states that in equilibrium, there will be equivalence between the CDS spread and the relevant bond spread. The bond spread being the difference between the sovereign bond yield and the risk-free rate. The risk-free rate can be chosen to be the relevant benchmark bond yield (in our case, the German bund yield) or the LIBOR/swap curve, [2]. We note that the benchmark bond yield is used by Delis and Mylonidis, [9] and by Haugh et al., [8]. As we wish to compare our results with those of Delis and Mylonidis, [9] we use the benchmark bond yield as the risk-free rate.

The data will be initially transformed into the log differentials of the spreads as is common practice in financial econometric analysis. The justification for this transformation is to remove the nonstationarity in the data caused by the inherent trend, [11]. In order to use linear stochastic modeling, the residuals of the model must be stationary and iid (i.e., independent and identically distributed). We will be analyzing this assumption, using linear and nonlinear tools. When applying the CRP methods, the log differentials are first normalized, this allows for the comparison of our results. The method of using log differentials is also applied by Crowley, [33] when analyzing synchronization of growth cycles across EU economies, but not by Basto and Caiado, [26] when comparing developed and emerging stock markets. Analyzing the untransformed spreads, allows the analysis of the nonstationary data series. Sprott [30] argues that using log differentials may be risky as the nonstationarity may be the interesting feature of the data. Be that as it may, we choose to analyze the log differentials as this allows us to compare our results with those in the financial econometrics literature.

4   Results

The methodological objective of the paper is to use linear and nonlinear tools to analyze the arbitrage-free parity theory for four euro zone peripheral CDS and bond spread markets. Initially we begin with an update of the Granger causality test (Table 1) which has been performed in many recent papers, [4, 7, 9]. Blanco et al., [7] analyze corporate CDS and bond markets and conclude that the CDS Granger causes changes in the bond spread in the majority of cases. Fontana and Scheicher, [4] review 10 sovereign euro-denominated markets and conclude that in

Table 1 Granger causality test, descriptive statistics and tests of equality

	Ireland		Spain		Portugal		Greece	
Granger causality (2 lags)	f stat	prob.	f stat	prob.	f stat	prob.	f stat	prob.
CDS does not GC Bond spreads	18.9337	1.00E-08	16.4654	1.00E-07	28.1461	2.00E-12	3.80911	0.0228
Bond spreads does not GC CDS	0.49478	0.61	0.84987	0.428	1.61915	0.199	25.3349	3.00E-11
Descriptive statistics	Bond spread	CDS	Bond spread	CDS	Bond spread	CDS	Bond spread	CDS
Mean	0.004215	0.002488	0.002954	0.002131	0.002463	0.002762	0.003485	0.002048
Median	0.000431	0	0.00412	0	-1.63E-04	0	0.00041	0
Maximum	0.721224	0.253736	0.633598	0.282618	0.791666	0.274167	0.225914	0.270875
Minimum	-0.598917	-0.331071	-0.804412	-0.417521	-0.87691	-0.590042	-0.208343	-0.230463
Std. Dev.	0.075341	0.052121	0.099887	0.059119	0.099858	0.062263	0.047616	0.036737
Skewness	0.297997	0.173972	-0.575238	-0.485028	-0.625593	-1.464205	0.313792	0.888554
Kurtosis	31.89112	8.782963	17.71942	9.352793	24.21428	19.21196	6.440282	17.77487
Jarque–Bera	18823.46	756.5823	4913.742	930.948	10180.07	6117.882	275.6712	4991.961
Probability	0	0	0	0	0	0	0	0
Observations	541	541	541	541	541	541	541	541
	value	prob.	value	prob.	value	prob.	value	prob.
t test for equality of mean	0.438344	0.6612	0.164838	0.8691	-0.059101	0.9529	-0.555457	0.5787
f test for equality of variance	2.089471	0	2.854678	0	2.572232	0	1.679971	0

half of the markets, bonds Granger cause CDS spreads and in the other half, CDS's Granger cause bond spreads. Interestingly, they find that for Ireland, Spain, Portugal, and Greece, CDS spreads Granger cause bond spreads. Delis and Mylonidis, [9] analyze sovereign CDS and bond spreads for Spain, Portugal, Greece and Italy and conclude that (when using a 250 day rolling Granger causality test), CDS spreads Granger cause bond spreads from 2007 onwards.

We apply the Granger causality test (Table 1). Our results agree with the previous research on sovereign euro zone markets [4, 9], that is, we show that CDSs Granger cause bond spreads.

Linear Gaussian models assume that the residuals are stationary, iid, and follow a Gaussian distribution. We will now review these assumptions by analyzing the univariate distributions of the log differentials of each of the CDS and bond spreads (Table 1). The descriptive statistics clearly show the non-Gaussian nature of each of the distributions (i.e., nonzero skew and excess kurtosis). The test of normality, the Jarque–Bera test, confirms this conclusion. By applying a t test for equality of means, comparing the mean of the CDS to the mean of the bond spread in each country, we see that the mean values appear to be statistically equivalent. Whereas by applying an f test to test the equality of the variance, we see that in all cases, the variance of the CDS spread is statistically different to that of the bond spread (Table 1). The bond spread has a higher variance than the CDS spread. This may be due to the fact that the bond market is exchange traded, whereas the CDS is over the counter (OTC), [1]. Or because the volume of trading in the sovereign bond market is significantly higher than that of the relatively new sovereign CDS market, [4]. We conclude that using sovereign CDS spreads to price bond markets will underestimate the variance and lead to mispricing of the assets. We also conclude that assuming a Gaussian distribution will underestimate the risk, as both markets exhibit excess kurtosis.

In general, linear stochastic models require the residuals to be iid, therefore we will run a simple linear regression around a constant to see if serial correlation in returns and variance are significant for CDS and bond spreads. If these correlations are shown to exist, they must be removed through correct linear modeling prior to applying a stochastic distribution to the data. We apply the LB Q statistic on the residuals and the squared residuals to test for serial correlation in returns and in variance, [11] (Table 2). We find evidence of serial correlation in returns in all markets except for the Portuguese bond spread and evidence of serial correlation in variance for all markets except for the Greek CDS. The linear correlation in returns is commonly removed in linear stochastic models but it is interesting to note that there is evidence of serial correlation in the variance also. ARGARCH model [36] can be applied to take account of this autocorrelation. If the autocorrelation is not removed, the residuals will not be iid, the estimates will be bias. This will increase the probability of a Type-1 error (i.e., false rejection of the Null Hypothesis).

We apply an ARGARCH model, to remove the linear dependencies in the data and then test the residuals for nonlinear dependencies through the application of the BDS test (Table 2). As in our previous study, where we analyzed the iTRaxx CDS index [13], we show evidence that there is remaining nonlinear dependence in the ARGARCH residuals of the sovereign CDS spreads. The bond spreads appear

Table 2 Tests for linear and nonlinear dependence

	Ireland		Spain		Portugal		Greece	
	Bond spread	CDS	Bond spread	CDS	Bond spread	CDS	Bond spread	CDS
LB Q stat - sc returns	8.8142	15.407	8.418	18.57	0.3845	29.699	7.5565	12.16
prob	0.012	0	0.015	0	0.825	0	2.30E-02	0.002
LBQ^2 stat - sc variance	23.399	13.746	91.108	15.432	42.344	10.349	23.235	4.0788
prob	0	0.001	0	0	0	0.006	0	0.13
BDS test - bond spread								
Dimension	z-Statistic	prob.	z-Statistic	prob.	z-Statistic	prob.	z-Statistic	prob.
2	0.675234	0.4995	−0.44643	0.6553	2.310341	0.0209	−0.38098	0.7032
3	−0.073508	0.9414	−0.417106	0.6766	1.578711	0.1144	0.409382	0.6823
4	−0.311324	0.7556	−0.405766	0.6849	1.601623	0.1092	0.8678	0.3855
5	−0.200452	0.8411	−0.206465	0.8364	1.69266	0.0905	1.188119	0.2348
6	0.636509	0.5244	−0.011419	0.9909	1.684624	0.0921	1.512419	0.1304
BDS test - CDS								
Dimension	z-Statistic	prob.	z-Statistic	prob.	z-Statistic	prob.	z-Statistic	prob.
2	6.774622	0	6.963198	0	8.7183	0	38.61742	0
3	6.419913	0	5.449211	0	8.838054	0	43.34863	0
4	6.240952	0	5.651973	0	8.577912	0	49.63063	0
5	5.751501	0	5.602257	0	8.715537	0	58.36041	0
6	5.62039	0	5.007999	0	8.482244	0	74.97942	0

to be independent. This result questions the use of equivalent modeling for CDS and bond spreads as well as the use of linear stochastic modeling for CDS spreads. It would appear from this analysis that linear Gaussian/stochastic modeling may not be appropriate in the analysis of the arbitrage-free parity theory for CDS and bond spreads. Thus we introduce a new methodology, that is, the CRPs and CRP measures.

In order to ensure consistency in our results, we applied the same embedding dimensions across all four CRP distance plots and the CRPs. The value of the embedding dimension m was chosen to be equal to 5, based on the false nearest neighbors method. The time delay, τ was kept equal to 1 as the data are discrete. Initially we analyze the CRP distance plots. These plots do not specify the threshold value; therefore, the plots indicate the closeness of the points through a color/graying scale (Fig. 2). The CDS market is represented on the x axis, and the bond spread on the y axis. As the eye moves vertically up the y axis, we are comparing one vector of CDS spread values to each vector of bond spread values. If the maximum distance between the two vectors is small, this will result in a dark spot. As the color lightens, this indicates the vectors are further and further away from each other.

We can see a white vertical band around May 2010, indicating a nonstationary period. The CDS values are far away from any of the bond spread values during the entire observation period. We can also see clear horizontal white lines between the 350 and 400 observation point on the y axis (this is just before and after May 2010). This indicates that the bond values at this time are far away from the CDS values during the observation period. These examples of nonstationarity would imply that the equivalence is nonmean reverting, [11]. In times of uncertainty and market intervention, the market values are no longer equivalent, suggesting caution when using pricing models, etc. We note that the Irish bailout in November appears to have no significant effect on the Irish market, although there is a white vertical band indicated in the Greek market. When comparing countries, we note that the number of white lines increase as we move from Ireland to Spain, to Portugal and is highest for Greece. This indicates the increasing nonstationarity as we move from one market to another, with a lack of equivalence being particularly noticeable in the Greek market.

We will now analyze the CRP plots, we choose a threshold of 0.7σ in all cases as this ensures consistency and comparability across markets (Fig. 3). If we allowed the threshold to vary, this may cause the differences to be due to the threshold value rather than actual market differences. By comparing these plots to Fig. 1, we can see that there appears to be some kind of checkerboard structure in the data. It does not appear to be random. This would imply that there is some deterministic relationship inherent in the two data sets and may possibly indicate attractors. We can again see white space around May 2010, but we also notice long horizontal and vertical lines. A vertical line indicates that the value in the bond spread remains trapped close to the value for the CDS spread for a period of time. A horizontal line indicates the same trapping for the CDS market. This characteristic would indicate equivalence continuing with a lag. We note that the equivalence is at times short lived. We also

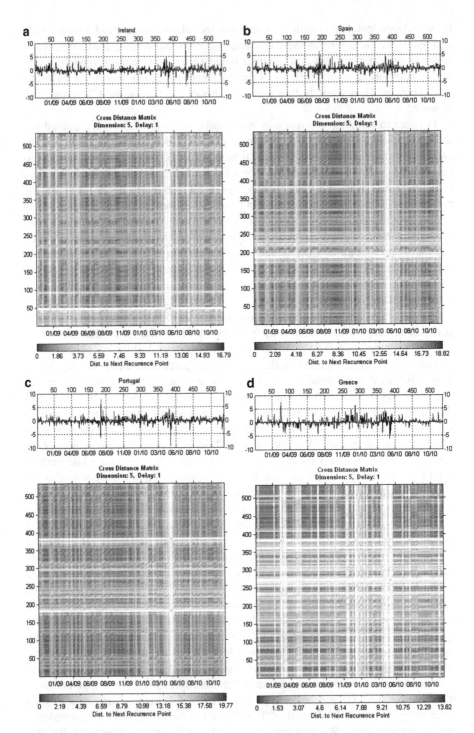

Fig. 2 CRP distance plots. (**a**) Ireland, *top left* (**b**) Spain, *top right* (**c**) Portugal, *bottom left* (**d**) Greece, *bottom right*

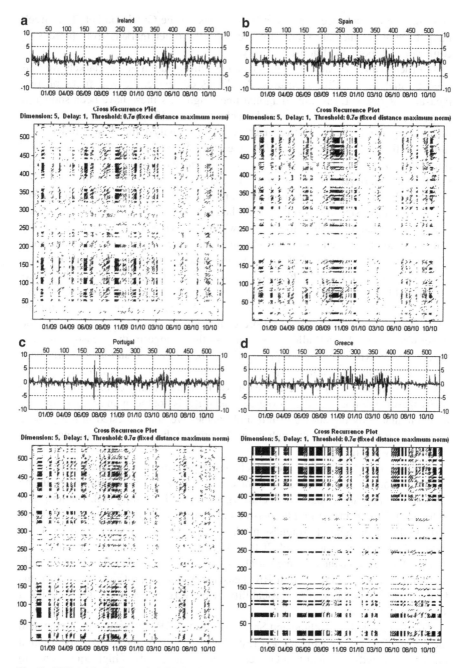

Fig. 3 CRP. (**a**) Ireland, *top left* (**b**) Spain, *top right* (**c**) Portugal, *bottom left* (**d**) Greece, *bottom right*

note that the equivalence appears to be strongest in the Greek market. The Greek market seems to fluctuate more violently than the other markets from periods of equivalence to periods of nonstationarity.

In order to analyze the convergence and synchronization in the markets we will need to analyze the CRP measures. As discussed above, we will focus on two measures, RR, and L (average diagonal line length), as these have been shown to be good measures of convergence and synchronization, [34]. RR allows us to assess the convergence between the two markets. We will analyze the change in RR over time by estimating a rolling value of the measure. We continue to use an embedding dimension, m, of 5 and time delay, τ, of 1. We reduce the threshold value to ensure that the number of recurrences is kept $\lesssim 1\%$. To do this we choose a value of 0.35σ. We keep the threshold fixed across all four countries to ensure comparability. We choose a window size of 60, as this will relate to 60 days, or a quarter of a year. This is a common length of time to analyze when reviewing financial markets (many statistics and announcements are made quarterly). We note that Basto and Caiado [26] chose a window size of 260 observations when analyzing equity markets, we found this window size to be too large, considering our sample size and the events we wished to analyze. Reducing the window size means the CRP measures reflect smaller scale dynamics, [24]. We choose a window step size of 1, which implies we will move one day forward before estimating the measure again. Using the windowed or epoch CRP allows us to assess the change in the measure through time, [24]. As RR measures the probability of occurrence of similar states in both systems, a rising RR implies a rising probability of equivalence; thus, analysis of RR through time gives us a relative expression of the validity of the arbitrage-free parity theory. We note that the absolute value of RR is not the main focus as this can be manipulated by the choice of threshold value. We are interested in the relative value across countries and through time. We present the RR values graphically for each country in Fig. 4.

The first four graphs indicate the changing probability of convergence in the CDS and bond markets across time for each country. We note similarities in the markets, with higher probability of convergence at the beginning of the period, with a downward trend in the middle period and a small recovery in the latter part of 2010 in the Spanish and Portuguese markets. At this time, we see a significant rise in the Greek market (from March 2010 onwards) with no such recovery in the Irish market. We also note the extreme variability in the Greek market. The relative scale is highlighted by placing all four RR measures in one graph (Fig. 4e). To test this further, we assess the average probability of convergence across markets using a statistical test of equality (Table 3).

The average RR for Ireland and Spain are similar, RR for Portugal is higher, and the RR for Greece is far higher. A t test illustrates this, as the hypothesis of equal means is rejected. It is clear that Greece swings from periods of high equivalence to periods of low equivalence, but overall the equivalence levels are higher for Greece than the other countries analyzed. To further illustrate the point, we applied the augmented Dicky Fuller test for a unit root on the RR measures. Acceptance of the null hypothesis suggests that a unit root exists in the dynamic probability of

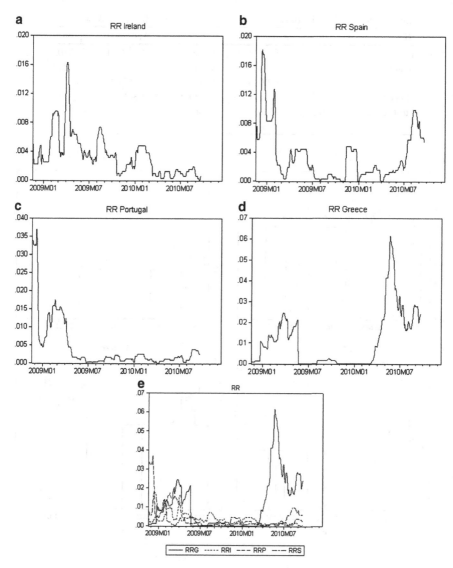

Fig. 4 Windowed RR measuring dynamic equivalence of the CDS and bond spread. (**a**) Ireland, *top left* (**b**) Spain, *top right* (**c**) Portugal, *middle left* (**d**) Greece, *middle right*. (**e**) All markets, last figure

convergence. We interpret this to imply that the probabilities are nonmean reverting. At a significance level of 1%, the null is rejected only for Portugal, all other markets exhibit a unit root. We note the particularly strong evidence of nonmean reversion in the Greek markets, this may suggest strong variability in arbitrage-trading activity in this market.

Table 3 RR: Probability of convergence, test of equality, estimate of skew

RR	mean	(s.e.)	ADF test	prob.
Ireland	0.003414	0.00013	-3.057761	0.0305
Spain	0.003355	0.000163	-3.162857	0.0229
Portugal	0.004321	0.000317	-5.368402	0
Greece	0.011139	0.000623	-1.860571	0.351
t test for equality of mean	Value	Prob		
	105.6099	0		

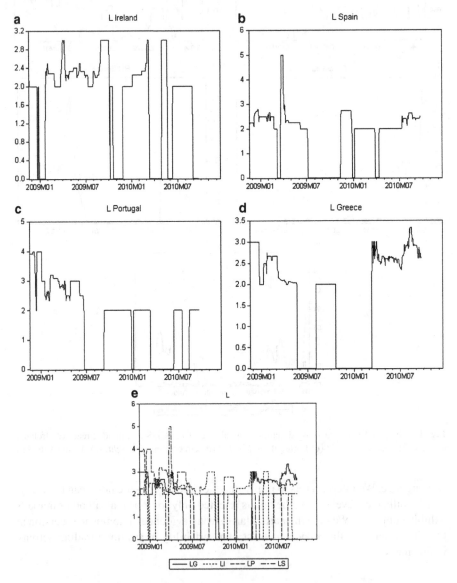

Fig. 5 Windowed L measuring common dynamics for the CDS and bond spread. (**a**) Ireland, *top left* (**b**) Spain, *top right* (**c**) Portugal, *middle left* (**d**) Greece, *middle right*. (**e**) All markets, last figure

Table 4 L: Synchronization/ common dynamics, test of equality

L	Mean	(s.e.)
Ireland	1.677955	0.054751
Spain	1.728222	0.046088
Portugal	1.715994	0.057406
Greece	1.750707	0.049023
t test for equality of mean	Value	Prob
	0.342761	0.7944

The second CRP measure we will analyze is L (the average diagonal line length). This will indicate the synchronization, or common dynamics across the CDS and bond spreads. Long diagonal lines indicate long periods of common dynamics. Thus a rising L implies rising synchronization between the two spreads and vice versa with a falling L. We are interested in the relative value across countries and through time. We present the L values graphically for each country in Fig. 5.

The first four graphs indicate the characteristics of the changing common dynamics between the CDS and bond spread in each of the countries. We note that each country appears to have a different pattern, with periods of high common dynamics and also periods where the dynamics/synchronization collapse to zero. Greece again is notable from March 2010 (i.e., prior to the bailout) we see a significant rise in the common dynamics. It is curious to note this recovery is prior to the bailout. Overall, if we place the four markets onto one graph (Fig. 5e), it appears as if the level of common dynamics is similar across the markets. We estimate the average value for L in each country and test the equality of the means in Table 4.

The t test clearly shows equality of mean estimates for common dynamics across the countries, suggesting that on average the level of synchronization is equivalent.

5 Conclusions

We have applied linear and nonlinear tools to test the arbitrage-free parity theory. The results of the linear Granger causality test concur with the results of previous literature, [4, 7, 9]. In general, sovereign CDS spreads are shown to Granger cause changes in sovereign bond spreads for the four countries under examination, that is, Ireland, Spain, Portugal, and Greece. This linear tool assumes that the assets follow a Gaussian distribution. The analysis of the descriptive statistics of the CDS and bond spreads in each of the four sovereign markets, indicate that both assets are non-Gaussian. Using a Gaussian model will lead to unreliable results. We also illustrate evidence of a statistically higher variance in the bond spread than in the CDS. This implies that assuming equivalent distributions for both assets will lead to poor fits and unreliable results.

We apply an ARGARCH model to each asset to remove the linear dependencies (in returns and in the variance) and note that nonlinear dependencies remain in the CDS data series. We conclude that linear stochastic models are not appropriate in the analysis of the CDS market. We also conclude that the application of linear Gaussian bivariate models for CDS and bond spread data will lead to unreliable results. We suggest nonlinear nonparametric tool should be applied instead.

We suggest the application of the nonlinear and nonparametric CRP and the CRP measures in order to evaluate the arbitrage-free pricing theory. The advantage of these nonlinear tools is that they can be applied to nonstationary, high dimensional data series. This methodology is unique in that we can examine the deterministic structure of the series without imposing a model.

Our results show that there are periods of nonstationarity in the equivalence between the two assets; particularly around the time of May 2010, which was a time of significant government intervention in the market. The existence of nonstationarities calls us to question the mean reversion assumption of the arbitrage-free parity theory.

In the CRPs, we show evidence of deterministic structures in the data and evidence that market is being trapped at certain levels. Equivalence being trapped for a period of time is a characteristic of a nonlinear system (not a periodic or a random system). This indicates that there may be an intermittent nonlinear relationship inherent in the two series. We note in both the CRP distance plots and the CRPs; that Greece appears to present stronger evidence of a deterministic structure as well as more nonstationarity. This suggests that country by country the markets behave differently over time.

The CRPs measures allow us to examine convergence and synchronization across the two data series, allowing a dynamical review of the arbitrage-free parity theory. We show that the probability of convergence (measured as RR) fluctuates widely across time and is not statistically equivalent across countries. In general, we conclude that convergence trends down initially (i.e., prior to the intervention) in all markets and trends upwards (i.e., after the intervention) in all but the Irish market. This trend in convergence indicates varying levels of arbitrage over time. We also note that Greece behaves differently with a significant rise in the convergence prior to the May 2010 intervention. By applying the ADF test, we conclude that the probability to converge is nonmean reverting (except in the Portuguese market). This result questions the assumption of a stable equilibrium, which is central to the arbitrage-free parity theory.

In the analysis of synchronization (measured as L) we note relatively stable dynamics across countries and through time (although there are periods when the common dynamics fall to zero). This indicates that there are common fundamental factors influencing the dynamics of the two markets. As evidence of synchronicity is found between the two markets, we suggest further analysis of the markets as a complex system.

There are a number of policy implications of this paper. Firstly, policy makers should be wary of results obtained from linear Gaussian models (for either of these two assets, be they univariate or bivariate). Secondly, they should be wary of results from linear stochastic models for CDS spreads. Thirdly, as the markets behave in an extreme and nonstationary manner around the time of government intervention; policy makers should be wary at this time of predictions made from stationary models. Fourthly, policy makers should be wary of the arbitrage-free parity theory; we show significant evidence to refute this theory. We note, in particular, the nonmean reversion of the equivalence between the asset classes. Policy makers cannot assume that market participants will take advantage of any arbitrage opportunity. In fact, we show evidence of trends in the probability to converge varying across countries. This indicates that specific markets go in and out of focus for market participants; that at times arbitrage increases and at times it falls. We show, in particular, that it rose for Greece prior to ECB intervention. This suggests that rising expectation of government intervention in a market leads to rising levels of arbitrage. By applying cross recurrence measures, policy makers will be able to calibrate and to supervise the probability of convergence between asset classes and the level of synchronization across countries. This knowledge and awareness of their influence on the markets could be particularly useful when drawing up new market intervention or bailout plans.[3]

References

1. Mengle, D.: Credit Derivatives: An Overview. In 2007 Financial Markets Conference. Federal Reserve Bank of Atlanta: International Swaps and Derivatives Association (2007)
2. Hull, J.C.: Options, Futures and Other Derivatives. 7th edn. Pearson Prentice Hall, New Jersey (2008)
3. Hull, J. C., White, A.: Valuing credit default swaps II: modeling default correlations. J. Derivatives 8(3), 12–22 (2001)
4. Fontana, A., Scheicher, M.: An analysis of the euro are sovereign CDS and their relation with government bonds. Working Paper Series, European Central Bank.(1271) (2010)
5. Duffie, D.: Presidential address: asset price dynamics with slow-moving capital. J. Finance 65, 1237–1267 (2010)
6. Mitchell, M., Pulvino, T.: Arbitrage crashes and the speed of capital. Working Paper Series, Mimeo (2011)
7. Blanco, R., Brennan, S., Marsh, I.W.: An empirical analysis of the dynamic relation between investment-grade bonds and credit default swaps. J. Finance 60(5), 2255–2281 (2005)
8. Haugh, D., Ollivaud, P., Turner, D.: What Drives Sovereign Risk Premiums? OECD Economics Department Working Papers(718) (2009)
9. Delis, M.D., Mylonidis, N.: The chicken or the egg? A note on the dynamic interrelation between government bond spreads and credit default swaps. Finance Res. Lett. In Press, Corrected Proof (2010)
10. Kantz, H., Schreiber, T.: Non Linear Time Series Analysis. 2nd edn. Cambridge University Press, Cambridge (2003)
11. Patterson, K.: An Introduction to Applied Econometrics. MacMillan, London (2000)
12. Granger, C.W.J.: Investigating causal relations by econometric models and cross-spectral methods. Econometrica 37(3), 424–438 (1969)

---

[3]We would like to acknowledge the use of the CRP toolbox 5.5 in the estimation of the CRPs and the CRP measures. This software was kindly given to use by Dr. Norbert Marwan of Potsdam University, see www.recurrence-plot.tk. We would also like to thank Dr. Stefan Schinkel, Dr. Denis O'Hora, Professor Stefan Thurner, and the referee for their valuable guidance and comments.

13. Moloney, K., Raghavendra, S.: Testing for nonlinear dependence in the Credit Default Swap Market. Economics Research International (Article ID 708704):12 (2011)
14. Brock, W.A., Dechert, W.D., LeBaron, B., Scheinkman, J.: A Test for Independence Based on the Correlation Dimension. Working papers Wisconsin Madison - Social Systems (1995)
15. Caporale, G.M., Ntantamis, C., Pantelidis, T., Pittis, N.: The BDS test as a test for the adequacy of a GARCH(1,1) specification: a monte carlo study. J. Financial Econometr. 3, 282–309 (2005)
16. de Lima, Pedro J.F.: On the robustness of nonlinearity tests to moment condition failure. J. Econometrics 76(1–2), 251–280 (1997)
17. Eckmann, J.P., Kamphorst S.O., Ruelle, D.: Recurrence plots of dynamical systems. Europhys. Lett. 4(9), 973–977 (1987)
18. Zbilut, J.P., Giuliani, A., Webber, C.L.: Detecting deterministic signals in exceptionally noisy environments using cross-recurrence quantification. Phys. Lett. A 246(1–2), 122–128 (1998)
19. Sauer, T., Yorke, J.A., Casdagli, M.: Embedology. J. Stat. Phys. 65(3–4), 579–616 (1991)
20. Takens, F.: Detecting strange attractors in turbulence. In: Rand, D., Young, L. (eds.) Dynamical Systems and Turbulence. Springer, Berlin (1981)
21. Kantz, H., Schreiber, T.: Non Linear Time Series Analysis, 2nd edn. Cambridge University Press, Cambridge (2003)
22. Marwan, N.: Encounters With Neighbours-Current Developments of Concepts Based On Recurrence Plots And Their Applications: PhD Thesis, University of Potsdam (2003)
23. Webber, C.L., Zbilut, J.P.: Recurrence quantification analysis of nonlinear dynamical systems. In: Riley, M.A., Van Orden, G., Arlington, C. (eds.) VATutorials in Contemporary Nonlinear Methods for the Behavioural Sciences. National Science Foundation (2005)
24. Marwan, N.: How to avoid potential pitfalls in recurrence plot based data analysis. In eprint arXiv:1007.2215: ARXIV (2010)
25. Schinkel, S., Dimigen, O., Marwan, N.: Selection of recurrence threshold for signal detection. Eur. Phys. J. Spec. Top. 164, 45–53 (2008)
26. Bastos, J.A., Caiado, J.: Recurrence quantification analysis of global stock markets. Phys. Stat. Mech. Appl. 390(7), 1315–1325 (2011)
27. Zbilut, J.P., Zaldivar-Comenges, J.M., Strozzi, F.: Recurrence quantification based Liapunov exponents for monitoring divergence in experimental data. Phys. Lett. A 297(3–4), 173–181 (2002)
28. Marwan, N.M., Romano, C., Thiel, M., Kurths, J.: Recurrence plots for the analysis of complex systems. Phys. Rep. 438(5–6), 237–329 (2007)
29. Eckmann, J.P., Kamphorst, S.O., Ruelle, D.: Recurrence plots of dynamical systems. Europhys. Lett. 4(9), 973–977 (1987)
30. Sprott, J.C.: Chaos and Time Series Analysis. 2nd edn. Oxford University Press, Oxford (2004)
31. Zbilut, J.P., Webber Jr. C.L.: Embeddings and delays as derived from quantification of recurrence plots. Phys. Lett. A. 171(3–4), 199–203 (1992)
32. Webber Jr, C.L., Zbilut, J.P.: Dynamical assessment of physiological systems and states using recurrence plots strategies. J. Appl. Physiol. 76(2), 965–973 (1994)
33. Crowley, P.M.: Analyzing convergence and synchronicity of business and growth cycles in the euro area using cross recurrence plots. Eur. Phys. J. Spec. Top. 164, 67–84 (2008)
34. Marwan, N., Kurths, J.: Nonlinear analysis of bivariate data with cross recurrence plots. Phys. Lett. A 302(5–6), 299–307 (2002)
35. Li, L., Mizrach, B.: Tail return analysis of Bear Stearns' credit default swaps. Econ. Model. 27, 1529–1536 (2010)
36. Bollerslev, T.: Generalized autoregressive conditional heteroskedasticity. J. Econometrics. 31, 307–327 (1986)

Index